Methods in
Molecular Biology
and Protein Chemistry

DATE DUE FOR R

Methods in Molecular Biology and Protein Chemistry

Cloning and Characterization of an Enterotoxin Subunit

Brenda D. Spangler
Montana State University, USA

JOHN WILEY & SONS LTD

Other Wiley Editorial Offices

John Wiley & Sons, Inc., 605 Third Avenue,
New York, NY 10158-0012, USA

Wiley-VCH Verlag GmbH, Pappelallee 3,
D-69469 Weinheim, Germany

John Wiley & Sons Australia Ltd, 33 Park Road, Milton,
Queensland 4064, Australia

John Wiley & Sons (Asia) Pte Ltd, 2 Clementi Loop #02-01,
Jin Xing Distripark, Singapore 129809

John Wiley & Sons (Canada) Ltd, 22 Worcester Road,
Rexdale, Ontario M9W 1L1, Canada

British Library Cataloguing in Publication Data

A catalogue record for this book is available from the British Library

ISBN 0 470 84360 8 (Paperback)

Typeset by Dobbie Typesetting Limited, Tavistock, Devon.
Printed and bound in Great Britain by Antony Rowe Ltd, Chippenham, Wiltshire.
This book is printed on acid-free paper responsibly manufactured from sustainable forestry,
in which at least two trees are planted for each one used for paper production.

Contents

Part 1: MOLECULAR BIOLOGY

Part 2: PROTEIN CHEMISTRY

Preface

An increasing number of students wish to enroll in a course that will guide them through up-to-date molecular biology and analytical biochemistry techniques and provide them with the tools and protocols that will enable them to plan and carry out an independent research project. To accomplish this objective The Chemistry and Biochemistry Department of Montana State University–Bozeman has, for the past five years offered a one-semester laboratory/lecture course in which students complete a research project, give one short library research talk on a topic associated with the project, write a mid-term research proposal and write a final research article suitable for publication. The project is expected to be an on-going one that can be expanded or modified to demonstrate the nature of scientific research. The material is suitable for Biochemistry, Biology and Microbiology senior undergraduate majors and minors, but is also relevant for Chemical Engineering, Chemistry, Biomedical Research, Animal or Plant Science majors as well. The laboratory text/manual is designed for a 3 credit-hour course, preferably done in two laboratory/lecture periods per week, since the molecular biology section in particular is more amenable to $2\frac{1}{2}$ hour work periods.

The project relies heavily on the primary literature in the field, and seeks to familiarize students with the necessity for using the library and the internet as research tools. Success in every experiment is not guaranteed: reagents may not work properly, the protocol may have been performed imperfectly, instruments may not work as expected. In other words, it is research as usual. A major goal of this course is to teach students, guided by the instructor, how to troubleshoot, recoup or re-do the experiment as necessary, and how to make progress toward a well-defined research goal. A back-up plan has been provided in the form of the plasmid which initiated the molecular biology section, so that, if the recombinant protein is not effectively expressed during the molecular biology section of the project, the protein chemistry can nevertheless go forward using protein expressed from the original plasmid.

Cloning and Characterization of the Binding Subunit of Escherichia coli Heat-labile Enterotoxin

The project mission is a systematic study of the structural factors which contribute to recognition of the membrane-bound receptors by the *Escherichia coli* heat-labile enterotoxin subunit known as LTB, responsible for specific binding to and entry into target cells. These relationships are designed to provide a database for manipulating the binding specificity of a naturally occurring multimeric protein delivery system and, as well, to define some of the structural features associated with molecular recognition.

The binding pentamers of both cholera toxin and the closely related *Escherichia coli* heat-labile enterotoxin are natural delivery vehicles which could be engineered to carry several types of macromolecules. Their immunogenicity and function make them useful as mucosal vaccine adjuvants in recombinant vaccines. Potentially, the well-documented retrograde transport capability of the binding pentamer in neurological research could also be exploited. Any project seeking to produce recombinant variants of a protein must use crystallographic methods as an analytical tool and this section of the project is therefore crucial. There have been a number of X-ray crystallographic studies as well as several recent attempts to exchange binding sites between cholera toxin and the closely homologous *Escherichia coli* heat labile enterotoxin by specific amino acid substitutions. The current project therefore includes an effort to crystallize the recombinant LTB and at least to demonstrate methods of protein structure determination.

The goal of the current project is to:

(1) perform and refine the protocols for obtaining and verifying recombinant LTB

(2) examine LTB binding to a panel of ganglioside receptor analogs

(3) obtain crystals of recombinant LTB for three-dimensional structural studies

(4) use the baseline data, combined with results from relevant literature, to design the primers for two new mutations.

Subsequent classes should be able to carry on by producing and assaying mutations and structural changes, then designing primers for the next class to test based on a chosen application.

One of the major contributions students make to this project is to define the ultimate research goal. Numerous recent literature reports describe a variety of applications for the binding subunit of *Escherichia coli* heat-labile enterotoxin. Students are encouraged to seek out these reports and to choose an application to improve, or, better yet, to devise a novel application for the recombinant protein. By so doing, they contribute to the ongoing nature of research. Modifications of protocol, for example, leaving an affinity tag sequence on the gene, can provide opportunities to extend the range of the research into new directions, using the framework of the initial cloning and characterization methods.

Finally, it may be helpful to inform new researchers and remind experienced ones that if we always knew what we were doing and the outcome was always predictable, it wouldn't be called 'research'. Put more formally, by Carlos Bustamonte (University of California-Berkeley), quoted in *Time* Magazine, 'Being a scientist means living on the borderline between your competence and your incompetence. If you always feel competent, you aren't doing your job'.

Acknowledgments

The author wishes to gratefully acknowledge the assistance of Dmitri Kazmin and Deborah Hyman who, as teaching assistants, tested the protocols, suggested changes, made back-up reagents, and helped to design the primers for this project; Dr. Michelle McGurl, now at the University of Montana, Missoula, who helped design the initial format of the course; Dr. Martin Lawrence and Mr. Ray Larsen, Montana State University, Bozeman, X-ray crystallographers who described and demonstrated X-ray crystallographic techniques to a succession of classes; The Chemistry and Biochemistry Department, Montana State University, Bozeman for supporting the course as the author defined it; technical advice from Bradley Elmore and Jerrod Einerwold, Montana State University, Bozeman; Paige Weiss and Barbara Morris, Novagen, Inc.; and William Jack, New England BioLabs, Inc. Plasmids pTZLT18 and pTZΔSA2B, together with much useful advice, were gifts from Dr. Witold Cieplak, Jr, Corixa Pharmaceuticals, Hamilton MT, and Dr. Ronald Messer, Rocky Mountain Laboratory, U. S. Public Health Service, National Institutes of Health, Hamilton, MT. Most especially, the author acknowledges the invaluable assistance of the several student classes who tested and critiqued the experiments and pointed out a multitude of errors, problems and glitches while gaining a gratifying level of expertise in research methods. This work is dedicated to my husband, Charles W. Spangler, without whose patience and encouragement it would not have been accomplished.

Brenda Spangler

Introduction and Background

History

The infectious disease known as cholera is caused by strains of the bacterium *Vibrio cholerae*. It is a severe diarrhea that has been epidemic in Southern India for probably thousands of years. It came to Europe by way of Turkey, the Russias and Poland in the late eighteenth and early nineteenth centuries and entered Western Europe by way of sailors who docked at Marseille, France in the 1830s. The disease moved from there to Paris, where it blossomed into a major epidemic, eventually spreading throughout mainland Europe, Britain, and North and South America. Cholera generated as much horror and revulsion among Europeans as bubonic plague had done previously. Cholera victims rapidly developed severe diarrhea and dehydration, resulting in toxic shock which left them shriveled, blue-black corpses, sometimes as rapidly as 4–6 h, depending on the size of the infective dose. The disease is a consequence of the biological activity of a bacterial protein toxin, cholera toxin (CT) secreted by the vibrio. Cholera remains a severe problem today in any area where poor sanitation allows it to spread. The strain responsible for the pandemic of the 1960s has been cultured and serves as the source of the commercially available toxin. Newer strains have developed in the 1990s that combine virulence with rapid replication, and are thus more dangerous than prior epidemic strains. Nevertheless, the mortality (death rate) is now 10–20% of those who develop the disease. Morbidity (disease development) is dependent on both dose (number of bacteria ingested) and the physical condition of those exposed. There is a native strain of *Vibrio cholerae* found in Gulf Coast cities of Louisiana and Texas that is responsible for occasional outbreaks in the USA.

Certain serotypes of *Escherichia coli* are well known as causes of diarrhea in swine, epidemics of swine 'cholera' having been common in the latter nineteenth and early twentieth centuries in the US. The secreted toxin responsible for swine 'cholera' was isolated from a strain of *E. coli* in 1969 (Gyles and Barnum, 1969). It was called *E. coli* heat-labile enterotoxin because it is rapidly inactivated by heat (heat-labile) and it is an enterotoxin, affecting the enteric (intestinal) system. It is also called LT, short for 'labile toxin'. During a major diarrheal outbreak in Calcutta, India, investigators (Gorbach *et al.*, 1971; Sack *et al.*, 1971) isolated from the bowels of victims a single, specific, serotype of *E. coli* rather than the expected organism, *Vibrio cholerae*. An epidemiological investigation of travellers' diarrhea reported in the British journal *The Lancet* in 1970 (Rowe, Taylor and Bettelheim, 1970) demonstrated that *E. coli* O148K/H28 was responsible for the diarrhea known by a multitude of common names: 'travellers' diarrhea', the 'Aden trot', 'Hong Kong dog', 'Delhi belly', and 'Gyppy tummy' in the Near and Far East, and 'tourista', the 'Aztec two-step' and 'Montezuma's revenge' in the New World. The pathogenic serotype secreted a toxin with activity that closely resembled that of CT from *Vibrio cholerae*. In fact, cholera antitoxin neutralizes the ability of LTp (the porcine enterotoxin) or LTh (the human-derived enterotoxin) to cause fluid accumulation in the isolated intestine of a rabbit. In enzyme-linked immunosorbant assays (ELISAs) anti-LTp or anti-LTh antibody reacts with cholera toxin and anti-CT antibody reacts with LT enterotoxin only slightly less well than the antibodies react with their respective cognate antigens.

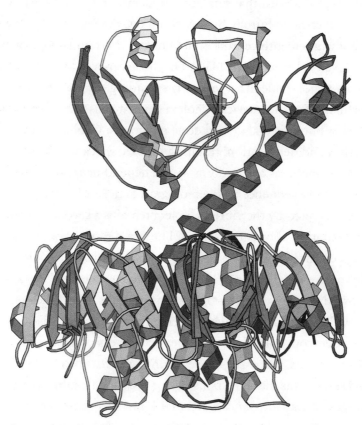

Figure I.1   This figure was created from the x-ray coordinates for *E. coli* enterotoxin (1LTS.pdb) obtained from the Protein Data Bank (www.rcsb.org/pdb) using Molscript (Kraulis, 1991). Image courtesy of Dr Melanie Rogers.

CT and the *E. coli* heat-labile enterotoxin LT are extremely homologous with one another, both structurally and functionally, suggesting a close evolutionary relationship. They are members of the structurally similar but functionally heterogeneous bacterial protein toxin family that includes pertussis toxin, diphtheria toxin, *Pseudomonas* exotoxin A, botulinum toxin, anthrax toxin, tetanus toxin, the Shiga and Shiga-like toxins and *Bacillus thuringiensis* toxin. These bacterial protein toxins are called A–B toxins because they consist of two functional parts, an 'A' component that has enzymatic function and a 'B' component that recognizes a target cell and serves as a delivery vehicle for the enzymatic portion of the toxin protein. Both CT and LT holotoxins have molecular weights approximately 85 620 Da, depending on source. Both CT and LT belong to the AB_5 family of ADP-ribosyl transferases. They consist of an A subunit, designated CTA or LTA, respectively ($M_r = 27\ 234$ Da), separated from the plane of a ring formed by five smaller subunits (11 600 Da each, CT; 11 800 Da, LT). The A subunit carries out the irreversible attachment of an ADP-ribose (derived from NAD^+) to the α subunit of a G-protein. The pentameric B subunits are designated CTB or LTB, respectively, and are responsible for recognizing and attaching to specific membrane-bound receptors (Spangler, 1992). The binding pentamer CTB or LTB is not itself cytotoxic.

Bacteria carrying the gene for CT or LT are ingested and pass to the small intestine of the mammalian host. After being released into the jejunum by the bacteria, the B pentamer of the toxin binds ganglioside GM_1 in the membrane of intestinal epithelial cells, or any other cell containing GM_1 in its membrane, such as neurons, brain cells and tear duct cells (Finkelstein *et al.*, 1974; King and Heyningen, 1973). Intoxication involves receptor-mediated endocytosis of the holotoxin and transportation to the Golgi. There the major portion of the A subunit, A1, is separated from the rest of the toxin by reduction of a disulfide bond. The A1 subunit is transported through the endoplasmic reticulum to the cytosol. Cholera is a more aggressive disease than 'Montezuma's revenge' (Cieplak, Jr *et al.*) because the vibrio preactivates CT by proteolytic cleavage of the peptide, whereas the *E. coli* does not. In addition, *E. coli* does not actively secrete its toxic products into the intestinal lumen. Instead, *E. coli* sequesters the toxin between the bacterial plasma membrane and the cell wall, necessitating cell lysis to release the toxin.

Enzyme Function

CT and LT transfer an ADP-ribose from NAD^+ (nicotinamide adenine dinucleotide) to the α subunit of a G-protein, Gs, associated with stimulation of the membrane-associated signal transduction enzyme adenylyl cyclase. G-proteins are trimeric, peripheral membrane proteins having three subunits designated α, β and γ. The α

subunit in the Gs (stimulatory) trimer is bound to a guanosine diphosphate (GDP) molecule. When nearby membrane-bound receptors are bound by an agonist, the $Gs\alpha-$GDP is phosphorylated and the resulting $Gs\alpha-$GTP (guanosine triphosphate) complex dissociates from the trimer. $Gs\alpha-$GTP interacts with adenylyl cyclase to form an active enzyme complex. The active ternary complex, consisting of adenylyl cyclase and $Gs\alpha-$GTP, converts ATP to cyclic AMP (cAMP). Cyclic AMP is known as a 'second messenger' because an external signal initiates its production and it subsequently activates kinases that initiate a cascade of internal cellular enzymatic activities leading to secretion of salt and water in secretory cells. After a short time, $Gs\alpha$ hydrolyses GTP to GDP. The $Gs\alpha-$GDP dissociates from adenylyl cyclase and the enzyme no longer actively catalyzes the formation of cAMP. When $GS\alpha$ is modified by the covalent addition of ADP-ribose it can no longer hydrolyze GTP to GDP. The ternary complex will therefore remain in an active configuration, continuously producing cAMP. High levels of cAMP inhibit sodium readsorption and activate chloride secretion (Field, 1993). This and other factors result in excess secretion of salt and water, at the rate of liters per hour, the severe 'rice-water' diarrhea and cramping characteristic of CT and LT intoxication.

Genetics

The A and B subunits of CT are encoded in *V. cholerae* on the *ctx*AB operon (Mekalanos *et al.*, 1983) which, together with 6 kilobases (kb) of flanking chromosomal DNA, was originally considered to be a genomic element. Recent studies, however, have demonstrated that the CTX virulence element corresponds to the genome of a lysogenic filamentous bacteriophage, CTXϕ, carried by the vibrio (Waldor and Mekalanos, 1996). This information was uncovered by finding that the *V. cholerae* El Tor strain CTX chromosomal element could be transferred to other strains of *V. cholerae* and recovered as a plasmid, a mobile genetic element. The CTX virulence element integrates into the El Tor strain chromosome but is extrachromosomal in other strains of *V. cholerae*. In *E. coli*, the genes *elt*A and *elt*B are plasmid-borne (Dallas, Gill and Falkow, 1979; So, Dallas and Falkow, 1978), leading to the suspicion that the gene may have been transferred to *E. coli* from *V. cholerae*. Sequence comparisons show that CTA differs only slightly from LTA (93% homology), but CTB differs from porcine LTB and human LTB more significantly (approximately 80% homology). It is likely that *E. coli* is the foreign host, since the protease that cleaves A1 from A2 is endogenous in *V. cholerae*, but is not present in *E. coli* O148. Most probably LTB diverged from CTB in its new *E. coli* host. The two toxin subunit genes are found either singly or the A subunit is found as a tandem repeat. They are transcribed into a single mRNA. Over-expression of the B subunit gene *elt*B accounts for the 5 to 1 excess of B subunit over A (Mekalanos *et al.*, 1983).

Binding Specificity

Despite their structural similarities, CTB and LTB differ significantly in their receptor-binding specificities (Fukuta *et al.*, 1988). Both toxins bind with high affinity to ganglioside GM_1, specifically to the galactose that terminates the oligosaccharide portion of the ganglioside (see Figure 22.2). CTB and LTB are also dependent on the sialic acid residue since neither binds asialo-GM_1 significantly. LT, however, binds with measurable affinity to other glycosphingolipids, including GM_2 and paragloboside. More recent work indicates that LT recognizes complex glycolipids containing chains of more than 30 saccharides (Backstrom *et al.*, 1997; Fukuta *et al.*, 1988). Such non-GM_1 receptors may account for much of the binding activity of LT. Structural studies of CTB and LTB in the presence and absence of various receptor analogs have provided insight into the structural factors involved in binding recognition (Merritt *et al.*, 1994a,b, 1995, 1997; Sixma *et al.*, 1991, 1992; Zhang *et al.*, 1995a,b). The receptor-binding site is located at the interface between monomer subunits of the pentamer and includes Trp88 on one monomer and Gly33 on the adjacent monomer. A recent structure-based exploration of the GM_1 binding sites of LT and CT (Minke *et al.*, 1999) revealed even more diversity in the panel of ligands recognized by LT. More than one amino acid may be involved in the differences between CTB and LTB binding specificity, an observation based on studies of CTB mutants containing amino acid substitutions that effectively turned CTB into LTB (Backstrom *et al.*, 1997).

More Background

Further details of structure and function can be found in the review (Spangler, 1992) and in Salyers and Whitt (2002), an extensive discussion of microbiological information. N. A. Williams and T. R. Hirst have written several recent articles reviewing the immunogenic properties of the enterotoxin B subunit that may be helpful for those who are interested in applications for modified LTB (Millar *et al.*, 2001; Williams, 2000).

References

Backstrom, M., Shahabi, V., Johansson, S., Tenneberg, S., Kjellberg, A., Miller-Podraza, H., Holmgren, J. and Lebens, M. (1997). Structural basis for differential receptor binding of cholera and *Escherichia coli* heat-labile toxins: influence of heterologous amino acid substitutions in the cholera toxin B-subunit. *Mol. Microbiol.* **24**, 489–497.

Cieplak, W. Jr, Messer, R. J., Konkel, M. and Grant, C. C. R. (1995). Role of potential endoplasmic reticulum retention sequence (RDEL) and the Golgi complex in the cytotonic activity of *Escherichia coli* heat-labile enterotoxin. *Mol. Microbiol.* **16**, 789–800.

Dallas, W. S., Gill, D. M. and Falkow, S. (1979). Cistrons encoding *Escherichia coli* heat-labile toxin. *J. Bacteriol.* **139**, 850–858.

Field, M. (1993). Intestinal electrolyte secretion: history of a paradigm. *Arch. Surg.* **128**, 273–275.

Finkelstein, R. A., Boesman, M., Neoh, S. H., LaRue, M. K. and Delaney, R. (1974). Dissociation and recombination of the subunits of the cholera enterotoxin (choleragen). *J. Immunol.* **113**, 145–150.

Fukuta, S., Twiddy, E. M., Magnani, J. L., Ginsburg, V. and Holmes, R. K. (1988). Comparison of the carbohydrate binding specificities of cholera toxin and *Escherichia coli* heat-labile enterotoxins LTH-I, LT-IIa, and LT-IIb. *Infect. Immun.* **56**, 1748–1753.

Gorbach, S. L., Banwell, J. G., Chattergee, B. D., Jacobs, B. and Sacks, R. B. (1971). Acute undifferentiated human diarrhea in the tropics. 1. alterations in intestinal microflora. *J. Clin. Invest.* **50**, 881–889.

Gyles, C. L. and Barnum, D. A. (1969). A heat-labile enterotoxin from strains of *Escherichia coli* enteropathogenic for pigs. *J. Infect. Dis.* **120**, 419–426.

King, C. A. and Heyningen, W. E. v. (1973). Deactivation of cholera toxin by a sialidase resistant monosialosylganglioside. *J. Infect. Dis.* **127**, 639–647.

Kraulis, P. J. (1991), MOLSCRIPT: A program to produce both detailed and schematic plots of protein structures. *J. of Appl. Crystallography* **24**, 946–950.

Mekalanos, J. J., Swartz, D. J., Pearson, G. D. N., Harford, N., Groyne, F. and deWilde, M. (1983). Cholera toxin genes: nucleotide sequence, deletion analysis and vaccine development. *Nature (Lond.)* **306**, 551–557.

Merritt, E. A., Sarfaty, S., Akker, F. v. d., L'Hoir, C., Martial, J. A. and Hol, W. G. J. (1994a). Crystal structure of cholera toxin B-pentamer bound to receptor GM$_1$ pentasaccharide. *Protein Sci.* **3**, 166–175.

Merritt, E. A., Sixma, T. K., Kalk, K. H., Zanten, B. A. M. v. and Hol, W. G. J. (1994b). Galactose binding site in *Escherichia coli heat-labile* enterotoxin (LT) and cholera toxin (CT). *Mol. Microbiol.* **13**, 745–753.

Merritt, E. A., Sarfaty, S., Chang, T.-t., Palmer, L. M., Jobling, M. G., Holmes, R. K. and Hol, W. G. J. (1995). Surprising leads for a cholera toxin receptor-binding antagonist: crystallographic studies of CTB mutants. *Structure* **3**, 561–570.

Merritt, E. A., Sarfaty, S., Jobling, M. G., Chang, T., Holmes, R. K., Hirst, T. R. and Hol, W. G. (1997). Structural studies of receptor binding by cholera toxin mutants. *Protein Sci.* **6**, 1516–1528.

Millar, D. G., Hirst, T. R. and Snider, D. P. (2001). *Escherichia coli* heat-labile enterotoxin subunit is a more potent mucosal adjuvant than its closely related homolog the B subunit of cholera toxin. *Infect. Immun.* **69**, 3476–3482.

Minke, W. E., Roach, C., Hol, W. G. J. and Verlinde, C. L. M. J. (1999). Structure-based exploration of the ganglioside GM$_1$ binding sites of *Escherichia coli* heat-labile enterotoxin and cholera toxin for the discovery of receptor antagonists. *Biochemistry* **38**, 5684–5692.

Rowe, B., Taylor, J. and Bettelheim, K. A. (1970). An investigation of travellers' diarrhoea. *Lancet* **1**, 1–5.

Sack, R. B., Gorbach, S. L., Banwell, J. G., Jacobs, B., Chattergee, B. D. and Mitra, R. C. (1971). Enterotoxigenic *Escherichia coli* isolated from patients with severe cholera-like disease. *J. Infect. Dis.* **123**, 378–385.

Salyers, A. A. and Whitt, D. D. (2002). *Bacterial Pathogenesis; a Molecular Approach*, 2nd edn. Washington, DC: ASM Press.

Sixma, T. K., Pronk, S. E., Kalk, K. H., Wartna, E. S., Zanten, B. A. M. v., Witholt, B. and Hol, W. G. J. (1991). Crystal structure of a cholera toxin-related heat-labile enterotoxin from E. coli. *Nature (Lond.)* **351**, 371–377.

Sixma, T. K., Pronk, S. E., Kalk, K. H., Zanten, B. A. M. v., Berghuis, A. M. and Hol, W. G. J. (1992). Lactose binding to heat-labile enterotoxin revealed by X-ray crystallography. *Nature (Lond.)* **355**, 561–564.

So, M., Dallas, W. S. and Falkow, S. (1978). Characterization of an *Escherichia coli* plasmid encoding for synthesis of heat-labile toxin: molecular cloning of the toxin determinant. *Infect. Immun.* **21**, 405–411.

Spangler, B. D. (1992). Structure and function of cholera toxin and the related *Escherichia coli* heat-labile enterotoxin. *Microbiol. Rev.* **56**, 622–647.

Waldor, M. K. and Mekalanos, J. J. (1996). Lysogenic conversion by a filamentous phage encoding cholera toxin. *Science* **272**, 1910–1914.

Williams, N. A. (2000). Immune modulation by the cholera toxin-like enterotoxin B subunits: from adjuvant to immunotherapeutic. *Int. J. Med. Microbiol.* **290**, 447–453.

Zhang, R.-G., Scott, D. L., Westbrook, M. L., Nance, S., Spangler, B. D., Shipley, G. G. and Westbrook, E. M. (1995a). The three-dimensional crystal structure of cholera toxin. *J. Mol. Biol.* **251**, 563–573.

Zhang, R.-G., Westbrook, M. L., Westbrook, E. M., Scott, D. L., Otowinowski, Z., Maulik, P. R., Reed, R. R. and Shipley, G. G. (1995b). The 2.4 Å crystal structure of cholera toxin B subunit pentamer: choleragenoid. *J. Mol. Biol.* **251**, 550–562.

EXPERIMENT 1

Instructions for Pipetman Use*

Introduction

Molecular biology manipulations are frequently done on a very small scale. Therefore it is essential for a researcher to be proficient in measuring and transferring very small volumes. Most solutions are measured by pipetting the required volumes with an adjustable pipetter, the most common type in molecular biology laboratories being a Rainin–Gilson Pipetman. Other manufacturers such as Brinkman Eppendorf or Oxford Finnpipette make similar devices. A Pipetman is a continuously adjustable digital pipet. Several models are available with specific ranges to increase accuracy. The Rainin–Gilson Pipetman is available as a P-20 for 0.5–20 µl; a P-200 for 20–200 µl; a P-1000 for 200–1000 µl (0.2–1 ml) and a P-5000 for 1000–5000 µl (1–5 µl). A P-2 is available for pipetting 1–5 µl, and a P-100 is also available. The P-20, P-200, and P-1000 are the most commonly used sizes. White or yellow tips fit P-100, P-200 and P-20 pipetters, and blue tips fit P-1000 pipetters. Larger pipetters generally have white tips.

Molecular biology techniques usually require sterile conditions. Thus, it is standard procedure to use a fresh, sterile pipet tip for each pipet manipulation, even if the volume and the solution are the same as the previous step. This also gives consistent pipetting results by eliminating the possibility of sample carry-over. The used tip is immediately discarded into a designated waste container. The extra care taken will ensure that bacterial cultures do not become contaminated and that techniques such as polymerase chain reaction (PCR), which may be highly sensitive to trace amounts of contaminants, give the desired result.

---

*Parts of this Experiment are reproduced by permission of Rainin Instrument Co. Inc.

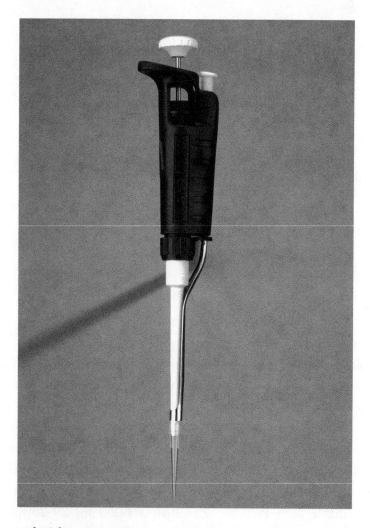

Figure 1.1   Diagram of a Gilson–Rainin Pipetman. Reproduced by permission of Rainin Instrument Co. Inc.

Materials and Methods

Materials

◆ Microbalance

◆ Pipetman pipetters, P-20, P-200, P-1000

◆ Tips

◆ 50% glycerol dyed with red food coloring, safranin or blue dextran

Method

(1) Hold the Pipetman in one hand. With the other hand turn the volume adjustment knob one-third of a revolution above the desired setting, e.g. 5 µl, then rotate slowly back until the required volume shows on the digital indicator. Always dial *down* to the desired volume. This is more accurate because of the way the Pipetman is designed. Volumes are read in microliters (µl) on the P-2, P-10, P-20, P-100 and P-200 Pipetman. Volumes are in ml on the P-1000 and larger Pipetman.

Note especially:

♦ The **P-20** has three numbers visible: the upper two, in black, are the integer number of microliters and the lower, red, number is the decimal. Thus, on a P-20, when you see 025 on the dial (02 in black, 5 in red) that means 2.5 µl. Observe what 2.5 µl look like in the tip. It is very tiny. On the other hand, a reading of 20 in black, 0 in red is 20 µl and would fill the tip about a quarter full.

♦ The **P-200** also shows three numbers, but all are black. Thus on a P-200, if you see 025, the pipetter would measure 25 µl. Observe what 25 µl look like in the tip. It would fill the tip about a third full. A reading of 200 would fill the tip.

♦ The **P-1000** uses the larger blue tips. Major divisions represent 10 µl. There are three numbers visible; the upper one is in red. A reading of 100 measures 1000 µl = 1 ml. A reading of 025 = 0.25 ml = 250 µl, while 090 = 0.9 ml = 900 µl, and 099 = 0.99 ml = 990 µl. There are graduations visible below the lowest number that can be used for finer measurement. For example, a reading of 025, pointed on the second graduation between 025 and 026 would measure 0.254 ml = 254 µl). All further volumes will be given in µl.

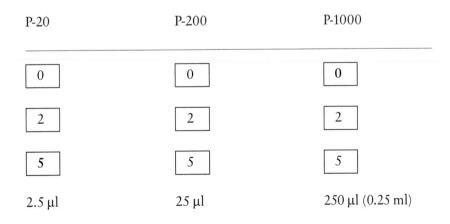

Figure 1.2   Digital volume indicators for the most common Pipetmen

(2) Attach a new disposable tip to the pipet shaft by inserting the shaft into the top of a tip situated in a tip rack. Press on firmly to ensure an airtight seal. Use white or yellow tips for the P-20 and P-200 pipets and blue tips for the P-1000.

(3) Depress the plunger to the *first stop* with the thumb. This part of the stroke is the calibrated volume displayed on the digital volume indicator.

(4) Holding the Pipetman vertically, immerse the disposable tip into the sample to the proper immersion depth: for very small volumes, immerse the tip just below the surface of the liquid. When removing large volumes with the P-1000 and P-5000, immerse the tip about 0.5 cm below the liquid surface and, since the volume decreases as it is drawn up, lower the tip accordingly.

(5) Allow the push-button to return slowly to the return position, using the edge of the thumb to slow the return. Never snap it up, or an air bubble will form, resulting in an inaccurate measurement. Wait until the full volume of sample is drawn up into the tip. Withdraw the tip from the sample and check for sample adhering to the tip. If this is a non-sterile sample, wipe the outside of the tip. If it is a sterile sample, try to minimize the amount of excess. Keep your eye on the tip: learn to recognize 1, 2, 5, 10 and 20 µl using the P-20; 50, 100, 150 and 200 µl using the P-200; and 250, 500 and 1000 µl using the P-1000 Pipetman.

(6) To dispense the sample, touch the tip against the side wall of the receiving vessel and depress the plunger to the *first stop*. Wait 1–2 s for the liquid to be expelled, then press the plunger to the bottom of the stroke to expel residual liquid. if you do not go to the second stop the measurement will be inaccurate.

(7) With the thumb keeping the plunger fully depressed, withdraw the Pipetman from the vessel, with the tip sliding along the wall of the vessel. *Do not let go of the plunger until the tip is clear of the liquid* or the sample will be sucked up again.

(8) Allow the plunger to return to the *up* position.

(9) Discard the tip using the tip ejector button. If the tip ejector has been removed to avoid contamination of the barrel, pull the tip off manually.

(10) Always use a fresh tip to avoid sample carry-over.

(11) Repeat steps 1–10 for 1, 10 and 20 µl (P-20), and for 50, 100 and 200 µl (P-200), and 500 and 1000 µl (P-1000). *Never turn the Pipetman beyond its maximum volume.*

Alternative mode: reverse pipetting

This method reduces error due to film retention but uses excess reagent:

(1) Mount the tip on the shaft.

(2) Depress the push-button to the *second stop.*

(3) Immerse the tip in liquid and return the push-button slowly to the *up* position.

(4) Wipe tip if necessary. Do not do so if sterility is an issue.

(5) To dispense, rest the end of the tip against the receiving vessel wall and depress the push-button to the *first stop.* Hold this position for a few seconds, long enough for the liquid to run out of the tip.

(6) Remove the tip from the receiving vessel *without blowing out the residual liquid.*

(7) Return the excess to the original sample, or discard.

(8) Discard used tip.

Hints

◆ Use a consistent rhythm while pipetting and consistent speed and smoothness when you press and depress the push-button. Pay attention to the stops. Pay close attention to the position of the tip and the size of the sample in it *before* dispensing the sample. This will avoid miscalculations and incorrect sample size.

◆ If there is an air bubble in the tip, dispense the sample into the original solution and pick it up again more carefully.

◆ Do not invert the Pipetman and *never* pipet without a tip. Drawing the liquid into the shaft will severely damage the Pipetman and it will have to be returned for expensive cleaning and re-calibration.

◆ Choose the right Pipetman for the job. For volumes between 1 µl and 20 ml use a P-20. For volumes between 10 and 200 µl (and preferably 20 µl as the lower limit) use a P-200. For volumes between 200 and 1000 µl use a P-1000.

Practice calibration

(1) Tare a small, covered vial containing 5 ml distilled water on a milligram balance or write down its weight to three places, four if possible.

(2) Pick up 5 μl of distilled water from a sample vial.

(3) Remove the cover from the tared vial and dispense the sample into the tared vial.

(4) Re-weigh the covered vial. The weight should change by 5 mg (0.005 g).

(5) Repeat steps (2)–(4) three times. PCR reactions and other common molecular biology techniques require the measurement of 1 μl or less with absolute accuracy. It is necessary to be very good at this.

(6) Repeat the calibration steps (2)–(4) three times using a P-200 and picking up 50 μl [the weight of the vial should change by 50 mg (0.050 g)].

Viscous Liquids

Try pipetting 5 and 50 μl of both a dyed 50% glycerol and a dyed 100% glycerol solution. It is difficult to pipet viscous solutions accurately with a Pipetman. Often displacement pipets are used for this purpose. Understanding the use of pipets with viscous liquids is important: DNA solutions are often viscous.

Some tricks for pipetting viscous samples:

(a) Draw the viscous liquid up slowly (30 s or more) keeping the level of the tip just below the surface of the sample.

(b) Cut off the edge of the tip to make a larger orifice (but note that accuracy may be compromised).

(c) Use a Pipetman that requires a larger tip to pipet the sample (eg use a P-1000 to pipet a 100 μl sample). This is equivalent to making a larger orifice on the tip, but should not compromise the accuracy of the sample.

Preparation for experiment

For each team:

♦ 1 box P-1000 blue tips (to be autoclaved)

♦ 1 box P-200 clear or natural tips (to be autoclaved)

♦ 1 roll 0.5 in. labeling tape, preferably a different color for each team

- Individual storage boxes for freezer and for refrigerator (clear plastic sandwich boxes and/or microcentrifuge tube racks)

- Permanent marking pen

- Drawer or storage tub for each team

- 50% glycerol containing food dye, safranin or blue dextrose

- 100% glycerol solution containing food dye, safranin or blue dextran

EXPERIMENT 2

Preparation of Plasmid Template DNA

Introduction

Plasmid

The plasmid pTZLT18 contains the DNA sequence for the entire LT operon, including the A and B portions of the enterotoxin amplified from the plasmid pEWD299. Plasmid pTZΔSA2B could also be used. It contains a shortened portion of the A gene sequence and the B subunit gene sequence. The complete LT operon originating in a pathogenic *E. coli* plasmid had been cloned into a new plasmid, pEWD299 (Dallas, Gill and Falkow, 1979). That plasmid was amplified and inserted into another plasmid to create the construct pTZLT18 in the expression vector *E. coli* DH5α (Cieplak *et al.*, 1995b; Grant, Messer and Cieplak, 1994). The original strain required that gene expression be induced by specific factors in the environment. The new plasmid construct carries its own promoter so the LTA and LTB genes cloned into it are constitutive (continuously expressed). The insert was prepared by gene amplification of the LT operon. The $3' \rightarrow 5'$ primer hybridized to the promoter region of the LT operon and included a *Bam*HI recognition site, while the $5' \rightarrow 3'$ primer hybridized to 125 nucleotides $3'$ of the termination codon of the B subunit gene and included a *Sal*1 recognition site at its $3'$ end. Further details of construction can be found in Grant, Messer and Cieplak (1994) and Cieplak *et al.* (1995b). The fragment containing the whole LT operon was 1400 bp (1.4 kb), and the plasmid into which it was ligated, pTZ18R (Pharmacia), was 2861 bp (2.9 kb). The total size of the construct (pTZLT18) is approximately 4300 bp (4.3 kb). In addition to this plasmid, one containing only the A2B gene fragment (pTZΔSA2B) was constructed (Cieplak *et al.*, 1995a). Either or both plasmids may be used as template for amplifying the LTB gene fragment. In practice, the A2B construct expresses only the B-subunit pentamer.

Preparation of Template by Alkaline Lysis Miniprep

There are several commercially available kits for plasmid purification, and the protocol can be done from scratch as well. Plasmid is isolated and purified by alkaline lysis (Birnboim, 1983). When 3–10 ml of culture are used it is called a 'miniprep'; 50–100 ml is a 'midiprep'. Large volumes of culture material require a 'maxiprep'.

The method, in brief, is:

(a) collect the cells by centrifugation to sediment (pellet) them; then discard the supernatant media;

(b) wash the cells by resuspending in buffer and re-sedimenting them;

(c) lyse (break open) the cells by addition of NaOH and sodium dodecyl sulfate (SDS);

(d) neutralize the solution with potassium acetate (KOAc), which also causes proteins and high molecular weight (i.e. chromosomal) DNA to co-precipitate, leaving plasmid DNA in solution. The precipitation is an effect of adding potassium (KOAc) to SDS creating a mixed-ion solution ($Na^+–K^+$ dodecyl sulfate);

(e) bind the plasmid DNA to a resin (if the kit used);

(f) wash the resin with high ionic strength buffer to remove unbound, soluble protein;

(g) elute the bound plasmid (stripping it from the resin) with a low-ionic strength buffer;

(h) re-precipitate the DNA with ethanol, to assure a clean, uncontaminated DNA sample.

The commercially available kits use spin columns to facilitate separation of plasmid DNA from bacterial components and bacterial chromosomal DNA. In general, recovery is excellent using these kits, because the plasmid DNA is bound to a spin column silica-gel membrane until elution by buffer, thus avoiding the problem of discarding an invisible DNA pellet along with the ethanol supernatant. A typical kit protocol is given first, then the detailed alkaline lysis protocol.

A miniprep of plasmid DNA can be accomplished using protocols outlined in Ausubel *et al.* (1999) or Sambrook and Russell (2000).

Materials and Methods

Materials: Spin miniprep*

♦ Microcentrifuge

♦ P-20 and P-200 Pipetmen, sterile white or yellow tips

♦ Plasmid purification kit (Wizard® Plus SV mini-prep, Promega Inc., Madison, WI catalog no. A1330; A QIAprep® Miniprep kit, QIAgen catalog no. 27104, may also be used, following manufacturer's instructions)

♦ Sterile 1.5 ml microcentrifuge tubes

♦ 3–5 ml culture of *E. coli* cells containing plasmid pTZLT18 in LB-Carb60 broth for each team

Method

Prepare lysate (optional, make glycerol stock)

(1) Place 1.5 ml of culture, in a sterile 1.5 ml microcentrifuge tube. Orient the tube centrifuge so the hinge faces inward. Centrifuge for 5 min to pellet the cells.

(2) Discard the supernatant and re-fill the tube with the all except 0.3–0.5 ml of the remaining culture, if making a glycerol stock to save for future use. Replace the tube in the centrifuge, hinge in, so the second pellet will sediment on top of the first. Centrifuge for 5 min.

(3) **Glycerol stock**   Add 75 ml 80% glycerol to the 0.3–0.5 ml culture remaining in the 15 ml centrifuge tube. Cap the tube tightly and store it at 4°C. The bacteria do not grow well in the absence of oxygen, and nor do they grow at 4°C. Glycerol stock can be frozen (preferably at −70°C) and stored for several months to be used as an inoculum later, should it be needed.

(4) Add 250 µl Wizard® Cell Resuspension Solution [50 mM Tris–HCl, pH 7.5; 10 mM ethylene diamine tetraacetic acid (EDTA); 100 µg/ml RNase A] to each sample and vortex well. The entire cell pellet must be resuspended. If necessary use a pipet to mix the solution.

*Miniprep method reproduced by permission of Promega Inc.

(5) Add 250 μl Wizard® Cell Lysis Solution (0.2 M NaOH; 1% SDS) and gently mix by inverting three or four times. Allow to incubate for 2 min at room temperature, until the solution appears clear. It will be very viscous if pipetted or stirred.

(6) Add 10 μl Wizard® Alkaline Protease Solution and gently mix by inverting three or four times. Incubate for 5 min at room temperature to allow the protease to break down cellular proteins.

(7) Add 350 μl Wizard® Neutralization Solution (4.09 M guanidine–HCl; 0.759 M potassium acetate; 2.12 M glacial acetic acid; final pH ≈ 4.2) and immediately mix by inverting three or four times. A white flocculent precipitate will form. The precipitate consists of denatured proteins associated with SDS.

(8) Centrifuge at full speed for 15 min at 10 000 *g* (or 10 min at 14 000 *g*).

Purify plasmid

(1) Transfer supernatant (∼850 μl cleared lysate) to a spin column provided with the kit. *Be very careful* not to disturb the large pellet. Decanting lysate into the column works well. If the solution becomes contaminated with precipitate, return it to a clean tube and centrifuge again.

(2) Centrifuge the spin column for 1–1.5 min at 10 000–14 000 *g*.

(3) Discard flow-through and re-insert the column in the receiver tube.

(4) Add 750 μl Wizard® Column Wash Solution containing ethanol (final concentration 60% ethanol; 60 mM potassium acetate; 10 mM Tris–HCl, pH 7.5) to the column.

(5) Centrifuge for 1–1.5 min.

(6) Discard the flow-through and re-insert the column in the receiver.

(7) Add 250 μl Wizard® Column Wash Solution.

(8) Centrifuge for 2–3 min until the membrane surface appears to be dry. Discard the supernatant.

(9) Place the column into a sterile 1.5 ml microcentrifuge tube. Add 100 μl nuclease-free water or 10 mM Tris–HCl, pH 8.5. If a higher concentration of DNA is preferred, use 50 μl of nuclease-free water or 10 mM Tris–HCl, pH 8.5.

(10) Centrifuge for 2 min to elute the plasmid DNA.

(11) Label the tube with identifying initials or number, the date, and pTZLT18. Store purified plasmid in the freezer.

DNA in water will be stable stored at $-20°C$ or below. DNA in Tris–EDTA (TE) buffer is stable at $4°C$. However, the EDTA chelates metals and will therefore inhibit the activity of some enzymes (that is how it inhibits bacterial growth). When DNA concentrations are low, requiring large volumes of sample to be added to an enzyme reaction mixture, the EDTA concentration is high enough to inhibit the desired reactions. Further purification by the ethanol precipitation described in steps (5)–(8) below can be performed using 2 vols 95% ethanol for the initial precipitation. For use in Experiment 3, however, this is not necessary.

Alternative Protocol

From Ausubel *et al.* (1999), pp. 1–22. See Ausubel *et al.* or Sambrook and Russell for buffer recipes.

Materials

◆ Microcentrifuge

◆ Sterile 1.5 ml microcentrifuge tubes

◆ 3 ml bacterial culture in LB-Carb50 broth (LB broth containing 50 µg/ml Carbenicillin)

◆ 50 mM glucose/25 mM Tris–HCl, pH 8/10 mM EDTA (GTE)

◆ 0.2 M NaOH/1% (w/v) SDS solution

◆ 5 M potassium acetate solution, pH 4.8

◆ 95 and 70% ethanol in water (v/v)

◆ 10 mM Tris–HCl pH 7.5/1 mM EDTA pH 8.0 (TE)

Method

(1) Place 1.5 ml of cell culture in a 1.5 ml microcentrifuge tube and orient the tube in the microcentrifuge so the hinge (or lip) faces out. Centrifuge 7000–10 000 g for 1 min. Repeat with the remaining culture.

(2) Discard the supernatant and resuspend the pellet in 100 µl GTE solution and let it sit for 5 min at room temperature.

(3) Add 200 µl NaOH/SDS solution, vortex and place on ice.

(4) Add 150 µl potassium acetate solution, vortex for 2 s, and place on ice for 5 min.

(5) Centrifuge for 7000–10 000 g for 3 min. Transfer 400 µl (0.4 ml) of supernatant to a clean tube. Add 800 µl (0.8 ml) of 95% ethanol and let it stand 2–5 min at room temperature.

(6) Centrifuge 7000–10 000 g for 3 min. Note the orientation of the tube and the expected position of the pellet. Remove supernatant very carefully, by sliding the pipet tip down the wall of the tube opposite to the pellet and slowly withdrawing the supernatant. Discard the supernatant.

(7) Wash the pellet with 1 ml 70% ethanol. To wash the pellet, carefully add 1 ml 70% ethanol to the emptied tube without disturbing the pellet. Centrifuge for 1 min making sure the tube is oriented exactly as it was in step 6. Remove and discard the supernatant. Dry the tube under vacuum or air dry for several minutes.

(8) Resuspend the pellet in 30 µl TE and store at 4 or −20°C. Alternatively, resuspend the pellet in 30 µl nuclease-free water and store at −20°C. DNA is stable stored at 4°C in TE. DNA to be used for enzymatic reactions and sequencing must be resuspended in water, since EDTA may inhibit enzymes that require Mg^{2+} as a cofactor.

A Note about Selective Media

Molecular biology media are formulated with antibiotics to select for only those cells that carry an antibiotic resistance gene provided by an introduced plasmid. Cells lacking the introduced plasmid will therefore die. Two common antibiotics used are Ampicillin and its analog Carbenicillin. Both Ampicillin and Carbenicillin are β-lactams that interfere with the final stage of bacterial cell wall synthesis. β-Lactams are cleaved by the enzyme β-lactamase, rendering the antibiotic ineffective. Antibiotic resistance is provided by the plasmid-borne gene *bla* encoding the enzyme β-lactamase. Kanamycin interferes with bacterial protein synthesis by binding to the 30s ribosomal subunit. Resistance to kanamycin is provided by the kan gene product which degrades the antibiotic. Plasmid vectors can carry one or both markers. However, when using the β-lactamase gene several precautions must be taken that are not necessary with kanamycin resistance. Ampicillin selection may be negated in cultures and on plates because the drug can be degraded by secreted β-lactamase, resulting in 'satellite' colonies that do not contain the resistance gene-carrying plasmid. These colonies can be seen growing at the edges of the original colony. In addition, a drop in pH that occurs normally during bacterial growth may degrade the antibiotic with similar results. Ampicillin plates degrade after 2 weeks. An Ampicillin analog, Carbenicillin, is more stable in low pH and plates made with this antibiotic can be kept for 3 months. Kanamycin resistance does not have these drawbacks.

Preparation for experiment

(1) *E. coli* culture: *E. coli* BL21 cells or DH5α cells infected with pTZLT18.
    Inoculate 30 ml LB-Carb50 broth with 50–100 µl glycerol stock. Incubate for 24 h at 37°C in a water-bath shaker. Plasmid DNA or DH5α cells carrying pTZLT18 can be obtained from the author (visit http://www.chemistry.montana.edu/bspangler.html).

(2) LB broth containing 50 µg/ml Carbenicillin, 5 g Bacto tryptone (pancreatic digest of casein, rich in Trp), 2.5 g yeast extract, 5 g NaCl, to 500 ml with distilled water. Autoclave mixture. Cool, then add 1 ml Carbenicillin stock.

(3) Carbenicillin stock: 25 mg Carbenicillin per ml distilled water. Keep frozen.

(4) It is useful to aliquot individual portions with a few (2–10) microliters extra for each solution used from the kit to avoid contamination of the bulk solution and simplify distribution.

(5) Tris buffers: It is common practice to make a 1 M stock of Tris base (12.1 g/100 ml distilled water). The pH is approximately 9.3 and the stock is unbuffered. The 1 M stock can be 0.2 micron filtered and kept at room temperature for several months. Tris buffers of various concentrations can be made from the stock. Dilute the 1 M stock with three-quarters of the amount of water needed to produce the desired final concentration (e.g. for a 10 mM Tris final concentration, dilute 1 ml of 1 M to 75 ml with distilled water). Adjust to the desired pH with 1 N HCl.
    If EDTA or azide (NaN_3) is to be added, they should be added before making the buffer up to its final volume since they are normally added from concentrated stock

solutions (see below) rather than as solids. Make up the buffer close to final volume. Add any solid reagents required in amounts that will provide the proper final concentrations in the final volume. Mix well and make up to final volume, e.g. for the 10 mM buffer made from 1 ml of 1 M stock, bring to 100 ml final volume with distilled water.

(6) EDTA 0.5 M stock solution: 18.6 g $Na_2EDTA \cdot 2H_2O$; 70 ml distilled water. Adjust pH to 8.0 with approximately 5 ml of 10 M NaOH while stirring. The EDTA is insoluble until the pH is near 8.

References

Ausubel, F. M., Brent, R., Kingston, R. E., Moore, D. D., Seidman, J. G., Smith, J. A. and Struhl, K. (1999). *Short Protocols in Molecular Biology*. New York: John Wiley.

Birnboim, H. C. (1983). Rapid alkaline extracton method for the isolation of plasmid DNA. *Meth. Enzymol.* **100**, 243–249.

Cieplak W. Jr, Mead, D. J., Messer, R. J. and Grant, C. C. R. (1995a). Site-directed mutagenic alteration of potential active-site residues of the A subunit of *Escherichia coli* heat-labile enterotoxin. *J. Biol. Chem.* **51**, 1–6.

Cieplak, W. Jr, Messer, R. J., Konkel, M. and Grant, C. C. R. (1995b). Role of potential endoplasmic reticulum retention sequence (RDEL) and the Golgi complex in the cytotonic activity of *Escherichia coli* heat-labile enterotoxin. *Mol. Microbiol.* **16**, 789–800.

Dallas, W. S. and Falkow, S. (1980). Amino acid sequence homology between cholera toxin and *Escherichia coli* heat-labile toxin. *Nature (Lond.)* **288**, 499–501.

Dallas, W. S., Gill, D. M. and Falkow, S. (1979). Cistrons encoding *Escherichia coli* heat-labile toxin. *J. Bacteriol.* **139**, 850–858.

Grant, C. C. R., Messer, R. J. and Cieplak, W. Jr. (1994). Role of trypsin-like cleavage at arginine 192 in the enzymatic and cytotonic activities of *Escherichia coli* heat-labile enterotoxin. *Infect. Immun.* **1994**, 4270–4278.

Sambrook, J. and Russell, D. W. (2000). *Molecular Cloning: a Laboratory Manual*. Cold Spring Harbor, NY: Cold Spring Harbor Laboratory Press.

So, M., Dallas, W. S. and Falkow, S. (1978). Characterization of an *Escherichia coli* plasmid encoding for synthesis of heat-labile toxin: molecular cloning of the toxin determinant. *Infect. Immun.* **21**, 405–411.

Spangler, B. D. (1992). Structure and function of cholera toxin and the related *Escherichia coli* heat-labile enterotoxin. *Microbiol. Rev.* **56**, 622–647.

EXPERIMENT 3

(a) Estimation of Plasmid DNA Concentration
(b) PCR Amplification of LTB Gene from Plasmid DNA

Introduction

Rapid estimation of DNA concentration

Before doing PCR it is advisable to determine how much template plasmid DNA is available. This procedure will verify the presence and concentration of DNA obtained in the Experiment 2 template preparation. While not precise, it allows verification of the presence of DNA in a very small sample, and estimation of the DNA concentration. The method is based on the fact that ethidium bromide, a small heterocyclic molecule, intercalates into the stacked base pairs of double-stranded DNA. Because ethidium bromide is a planar tricyclic ring system, it lies perpendicular to the helical axis and its peripheral phenyl and ethyl rings project into the major groove of the DNA helix. The bound dye fluoresces 20–25-fold more strongly than the free dye in solution, emitting in the red-orange region when UV (ultraviolet) light at 302 nm is absorbed. Thus, the ethidium bromide serves as an indicator for DNA either in solution or immobilized in a gel. It will also bind to RNA (where there is intrastrand base pairing) and to denatured or single-stranded DNA, although less well in both cases (Sambrook and Russell, 2000, A9.3). When incorporated into a gel, ethidium bromide serves as a 'stain' to detect DNA fragments. It should be noted, however, that the presence of the dye reduces the mobility of linear double strands somewhat. Due to its intercalation into the helix, ethidium bromide stretches the helix open slightly, resulting in some nicking of the DNA strands. It is also a potential mutagen if the stretched helix causes DNA-dependent RNA

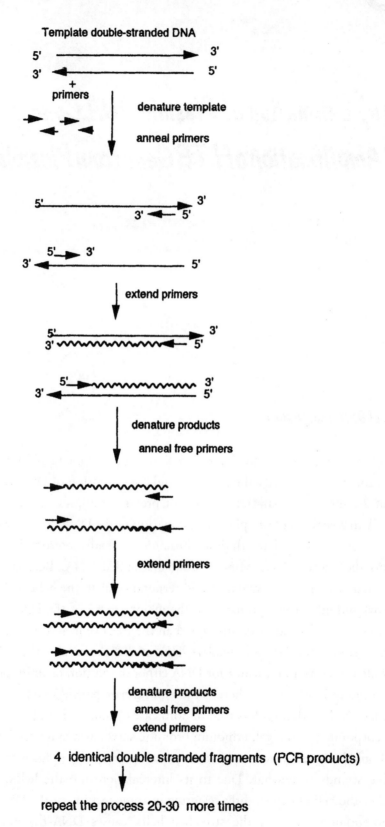

Figure 3.1   PCR scheme

polymerase to skip a base during transcription. Although it is a very weak mutagen in the Ames Mutagen Test, *wear gloves* when using ethidium bromide; *wear goggles* when viewing ethidium bromide preparations in ultraviolet light. To dispose of ethidium bromide in buffer, collect the solution in a clearly marked waste container and notify the proper waste disposal personnel for removal. Gels containing ethidium bromide should be wrapped in plastic cling film and placed in a ziploc bag with a piece of paper towel. They can then be disposed of with normal waste since most of the ethidium bromide is in the excised DNA slice.

Polymerase chain reaction

The PCR is a method for making many copies of (amplifying) a specific DNA sequence. It is an *in vitro* method devised by K. Mullis (Mullis *et al.*, 1986) based on the process by which DNA is made *in vivo*. Mullis wrote a popular article (Mullis, 1990) describing the background for the technique and how he devised it during a moonlit drive up the California coast.

Briefly, PCR consists of three basic reaction steps that are repeated many times (Figure 3.2):

(1) *denaturation*, in which single-stranded DNA is produced by heating the double-stranded DNA template to approximately 95°C;

(2) *annealing*, in which the section of the template strand to be amplified (gene of interest or DNA fragment) is 'chosen' by annealing, at approximately 50°C, a short $3' \rightarrow 5'$ primer piece of DNA (reverse primer) complementary to the $5' \rightarrow 3'$ template strand containing the last nucleotides of the section to be amplified, and a $5' \rightarrow 3'$ primer DNA (forward primer) that is complementary to the $3' \rightarrow 5'$ template strand containing the beginning of the gene of interest (see scheme, Figure 3.2); thus the gene of interest is bracketed by the primers, each of which have open $3'$–OH groups that can initiate the next step of the reaction;

(3) *chain elongation* or *extension* in which DNA polymerase adds deoxynucleotides (dNTPs) to each primer to make copies of both strands of template DNA.

The three steps are repeated, and for each repeat, n cycles, the number of copies doubles, increasing the copy number exponentially (2^n); (see Figure 3.1). It is clear that 20–30 cycles of amplification result in a huge increase in the amount of PCR product. Figures 3.1 and 3.2 show only one template molecule. In fact, there are hundreds of original templates and many thousands of primer molecules in the reaction mix.

Cycle number	Number of dsDNA products
1	0
2	0
3	2
4	4
5	8
6	16
7	32
8	64
9	128
10	256
11	512
12	1024
13	2048
14	4096
15	8192
16	16 384
17	32 768
18	65 536
19	131 072
20	262 144
21	524 288
22	1 048 576
23	2 097 152
24	4 194 304
25	8 388 608
26	16 777 216
27	33 544 432
28	67 108 864
29	134 217 728
30	268 435 456
31	536 870 912
32	1 073 741 824

Figure 3.2   Amplification of dsDNA fragments

Choosing the correct annealing temperature ensures that nonspecific annealing will be minimized and only precise complementarity will occur. If the temperature is too low, there is the possibility of non-specific annealing. However, if the temperature is too high, even specific annealing will be difficult and product yield will be reduced. If a primer contains a mutation, the annealing temperature must be adjusted downward to allow annealing of the strand that includes the mismatch. The beauty of the technique is that each subsequent short unit becomes a template itself for further amplification, so mismatches due to the inserted mutation become moot. The original template molecules are swamped out by the huge concentration of short PCR fragments that are used as subsequent templates.

The efficiency and precision of the technique require the use of high temperatures for the denaturation step. Thermostable polymerase enzymes that can operate repeatedly at temperatures above the melting point of DNA are therefore crucial for successful PCR. Commercially available polymerases are obtained from thermophilic bacteria. The first one used, *Taq* DNA polymerase was obtained from *Thermus aquaticus*, a pink filamentous bacterium found in 1967 growing in the outflow of a hot spring in Yellowstone Park, Wyoming (Brock, 1994). *Vent* DNA polymerase is obtained from pyrococcal species of bacteria found near a deep-sea vent, as are several others, including *Pfu* DNA polymerase, *Pfx* DNA polymerase and *Pwo* DNA polymerase, all from *Pyrococcus* species found near deep-sea vents. Each of these polymerases has characteristic activities such as $3' \rightarrow 5'$ exonuclease activity (proofreading ability); error rate (mismatch rate); stability at high temperature; extension rate (dNTP added/min); processivity (length of daughter strand before polymerase disengages); and particular specificity (*Taq* for example adds an extra A to the $3'$ end of PCR products, *Vent* and *Pwo* produce PCR products with blunt ends). *Pfx* DNA polymerase, which was originally obtained from *Pyrococcus* species but is now cloned in *E. coli* and supplied in a recombinant form, will be used. *Pfx* DNA polymerase has a molecular weight of about 90 kDa and is a highly processive $5' \rightarrow 3'$ DNA polymerase with proofreading activity. It is very thermostable, with a half-life of 2 h at 100°C, compared to *Taq*, which has a 5 min half-life at that temperature. Because *Pfx* has proofreading ability, the fidelity of replication is very high.

The PCR reaction is performed in a thermal cycler that can be programmed to change temperatures rapidly for each step. Typically, there is an initial denaturation step lasting 3–4 min to denature any contaminating protein, apart from the polymerase which is not temperature sensitive, followed by a short denaturation step (1 min) at 94–96°C, then temperature reduction to 50–65°C for annealing and 0.5–4 min at 65–70°C for extension. The next cycle requires ramping the temperature back to 94–96°C to repeat the series of steps. Temperature optimization, amount of Mg^{2+} in the buffer, amount of EDTA, concentration of template DNA and extension times are all important variables that must be considered, particularly when using primers that may anneal to other parts of the template, producing an undesired PCR product. $MgSO_4$ is required for the reaction. It is generally believed to stabilize the phosphate functional groups on the nucleotides.

The target gene

We have begun this project by preparing DNA for use as a PCR template, from which the gene will be amplified for the B subunit of LT. The LT genetic determinant was

5' GTTGACATATATA ACAGA ATTCGGGATGA ATT ATG A AT A AA GTA A AA TGT

                                                   Met Asn Lys Val Lys Cys

5' TAT GTT TTA TTT ACG GCG TTA CTA TCC TCT CTA TAT GCA CAC GGA GCT

    Tyr Val Leu Phe Thr Ala Leu Leu Ser Ser Leu Tyr Ala His Gly $\underline{Ala_1}$

5' CCC CAG ACT ATT ACA CAA CTA TGT TCG GAA TAT CGC AAC ACA CAA

    Pro Gln Thr Ile Thr Glu Leu Cys Ser_{10} Glu Tyr Arg Asn Thr Gln

5' ATA TAT ACG ATA AAT GAC AAG ATA CTA TCA TAT ACG GAA TCG ATG

    Ile Tyr Thr Ile_{20} Asn Asp Lys Ile Leu Ser Tyr Thr Glu Ser_{30} Met

5' GCA GGC AAA AGA GAA ATG GTT ATC ATT ACA TTT ATG AGC GGC GAA

    Ala Gly Lys Arg Glu Met Val Ile Ile_{40} Thr Phe Met Ser Gly Glu

5' ACA TTT CAG GTC GAA GTC CCG GGC AGT CAA CAT ATA GAC TGC CAG

    Thr Phe Gln Val_{50} Glu Val Pro Gly Ser Gln His Ile Asp Ser_{60} Gln

5' AAA AAA GCC ATT GAA AGG ATG AAG GAC ACA TTA AGA ATC ACA TAT

    Lys Lys Ala Ile Glu Arg Met Lys Asp_{70} Thr Leu Arg Ile Thr Tyr

5' CTG ACC GAG ACC AAA ATT GAT AAA TTA TGT GTA TGG AAC AAT AAA

    Leu Thr Glu Thr_{80} Lys Ile Asp Lys Leu Cys Val Trp Asn Asn_{90} Lys

5' ACC CCC AAT TCA ATT GCG GCA ATC AGT ATG AAA AAC TAG

    Thr Pro Asn Ser Ile Ala Ala Ile Ser_{100} Met Lys Asn STOP

Figure 3.3   DNA sequence for the LTB monomer (Dallas and Falkow, 1980). The sequence represents the A2B portion of the enterotoxin gene. The A2 sequence overlaps the B subunit. For additional information, see Spangler (1992)

cloned from a plasmid found in a porcine strain of *E. coli* by So, Dallas and Falkow (1978). Subsequently the genes, named *elt*A (encoding LTA) and *elt*B (encoding LTB; Dallas, Gill and Falkow, 1979) were identified. The LTB nucleotide sequence was translated into an amino acid sequence which showed a 21 amino acid leader sequence for secretion through the plasma membrane, and 103 amino acid residues comparable to cholera toxin.

Primers:

LTB-F(NcoI) 5′ ACTGC/CC-<u>ATGG</u>/ATAAAGTAAAATGTTATGT 3′ anneals to the 3′ → 5′ template strand at the beginning of the gene, and contains an *Nco*I cleavage site / CCATGG/. Starting with ATG (Met), the amino acid sequence is MNKVKCY. Note that the inclusion of the *Nco*I site necessitates changing the second amino acid from Asn (AAT) to Asp (GAT). Since the site precedes the B subunit sequence, this modification should not affect the B subunit structure.

   LTB-R(BamHI) 5′ TGAGC/GGATCC/-<u>CTAG</u>TTTTTCATACTGATTG 3′ anneals to the 5′ → 3′ template strand at the end of the gene, and contains the complement (/GGATCC/) of the *Bam*HI site /CCTAGG/ and the underlined complement of the stop codon.

   Amplification proceeds according to the scheme shown in Figure 3.2.

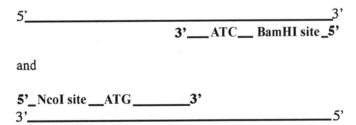

Figure 3.4   Positioning of primers on template DNA′ under 3′ → 5′

Primer design

The primers have been designed to contain a specific restriction endonuclease site at their 5′ end to facilitate insertion into the cloning vector. The forward primer that will be used, LTB-F(NcoI), contains an *Nco*I site. The coding portion begins with the ATG (Met) start codon that precedes the LTB sequence, and a 'GC clamp' at the 5′ end to facilitate complementary strand formation during PCR. The reverse primer, LTB-R(*Bam*HI), contains the complementary *Bam*HI restriction endonuclease site and the stop codon complementary sequence. The two 3′ ends of the primers are not complementary to one another, to prevent formation of primer–dimers. It is also necessary to provide at least five nucleotides preceding and after the restriction endonuclease site to provide a stable binding site, a 'perch' for the enzyme so it can interact with the phosphate backbone of the DNA (W. Jack, NEBioLabs, personal communication). The restriction endonuclease target sites that will appear in the PCR product DNA have been chosen carefully, so that both restriction enzymes are able to operate in the same buffer system. It will then be possible to use a single mixed digestion to prepare the insert for ligation into its vector.

Alternative forward primers are possible. The choice of a forward primer changes the method used to isolate the expressed protein. The pET28b vector used in these experiments contains sites for adding a poly-Histidine amino acid sequence (His-Tag) to the expressed protein. The amino-terminal His-Tag, if left in the vector by using an *Nde*I cleavage site from an NdeI primer, enables use of an immobilized Ni^+ affinity column to isolate expressed protein. The poly-His sequence is cleaved enzymatically from the expressed protein after isolation and the cleavage products applied to another column to purify LTB from the reaction mixture. However, isolation and purification of wild-type protein can be simplified by removing the poly-His site from the DNA sequence using the *Nco*I insertion position (*Nco*I primer). In that case, poly-His will not be part of the expressed protein and an immobilized galactose (a receptor analog) affinity column can be used to isolate and purify the protein in a single step. Note that a C-terminal His-tag is also part of the pET 28b sequence, but the LTB gene's stop codon precedes it, so it will not be a part of the protein. The pET 28b vector carries a variety of other cleavage sites and tags as well as a ribosomal binding site and a T-7 promoter sequence that facilitate protein expression.

Planned mutations of the LTB gene which result in a modified receptor-binding site would make it necessary to use the His-Tag for isolation of mutated protein, since mutations in the binding site might be expected to modify the ability of the mutated protein to bind an immobilized galactose (receptor analog) affinity column.

In this series of experiments the *Nco*I site is used to obtain LTB native protein directly for isolation on a galactose affinity column.

Materials and Methods

Materials

Rapid estimation of DNA concentration using ethidium bromide

♦   UV transilluminator, goggles, gloves

♦   P-20 Pipetman and sterile white or yellow tips

- DNA standards: 0, 1, 2.5, 5, 7.5, 10 and 20 ng/µl in nuclease-free water or 10 mM Tris–HCl, pH 8.5

- Ethidium bromide test solution: 1 ng ethidium bromide/µl nuclease-free water or 10 mM Tris–HCl, pH 8.5

- Purified plasmid DNA samples from Experiment 2

- Plastic clingfilm

PCR

- Microcentrifuge with 0.5 ml tube rotor (1.5 ml tubes with lids cut off can be used as adapters in the 1.5 ml rotor)

- Thermal cycler

- P-20 Pipetman and sterile white or yellow tips

- Sterile 200 µl dome-top PCR tubes

- Tube holder trays

- Ice and individual ice containers

- Template DNA as backup (pTZLT18, $\sim$0.4 ng/µl)

- DNA sample from Experiment 2, $\sim$0.4 ng/µl

- Primers: 4 µM LTB-F (NcoI); 4 µM LTB-R (BamHI)

- *Pfx* DNA polymerase (Invitrogen)

- 10$\times$ *Pfx* PCR buffer (supplied with the enzyme)

- 5 mM dNTP mix diluted in 1$\times$ PCR buffer from 10 or 25 mM stock to contain 5 mM each of dATP, dCTP, dGTP and dTTP

- 25 mM $MgSO_4$ (supplied with enzyme)

- Sterile (nuclease-free) HPLC-grade or deionized distilled water

Method

Rapid estimation of DNA concentrations using ethidium bromide. Ausubel et al. (1999), pp. 2–24.

This method is reasonably sensitive and is useful for estimating DNA concentrations smaller than 20 µg/ml (20 ng/µl) that may be difficult to quantitate with a spectro-photometer. Since the fluorescence intensity cannot be discriminated visually for concentrations above 20 µg/ml, the spectrophotometric method (Experiment 6) must be used for higher concentrations.

(1) Prepare DNA standards in 10 mM Tris, pH 8.5 by direct dilution. *Wear gloves when using ethidium bromide* [DNA Standards: 0, 1, 2.5, 5, 7.5, 10 and 20 and (optional) 50 µg/ml; the standards can also be expressed as 0, 1, 2.5, 5, 7.5, 10 and 20 ng/µl].

(2) For the standard curve, use a P-20 Pipetman to make a row of seven or eight spots, 4 µl each, of a well vortexed 1 µg/ml ethidium bromide test solution on a piece of plastic film placed on a UV transilluminator. Add 4 µl of each standard solution, one standard solution per spot. Mix gently by pipetting up and down two or three times. *Use a fresh tip for each standard.*

(3) Place a 4 µl spot of 1 µg/ml ethidium bromide test solution under the row of standards and add 4 µl of sample DNA. Mix gently.

(4) Turn on the UV transilluminator and estimate the unknown DNA concentration by visual comparison to the fluorescence of the standards. *Use goggles when viewing; no. 30 sunscreen is also advised.*

(5) In general, a 3 ml miniprep yields at least 20–50 ng/µl in 30–50 µl. Based on the estimated concentration of DNA obtained in Experiment 2, dilute a small amount to approximately 0.4 ng DNA/µl nuclease-free water for the PCR template. A midiprep in which 300 ml of culture is reduced to 300 µl of DNA will yield a 10-fold higher concentration of DNA.

Optimal template concentrations may vary depending on the template. Sambrook and Russell (2001, Section 8.20) suggest a plasmid DNA template stock containing 1–5 ng/µl, and a template concentration of 1pg/µl per ml of reaction mix. Others use higher concentrations. It is advisable to do several template dilutions in a preliminary test to determine best yield. Previous experiments determined, for example, that 80 pg pTZLT18 reaction mixture results in an excellent yield of amplification product. If the sample is diluted to a different stock concentration, e.g. 1 ng/µl, the amount used in the reaction mixture must be adjusted accordingly.

The pTZLT18 used as back-up template in this experiment is from a midiprep that yielded 200 ng/µl. It is diluted 1:500 for use in the PCR reaction. Calculate the concentration of stock DNA and verify that 10 µl of 1:500 pTZLT18 stock will give 80 pg template DNA/µl assuming a 50 µl reaction mixture.

PCR considerations

Primer design programs have an option for calculating the annealing temperature, which is an important consideration for successful PCR. Although the usual annealing temperature is 50 or even 55°C, the choice of 40°C for the annealing temperature is based on the primer sequences. The forward primer has a run of 4 As and the reverse has a run of 5 Ts. Neither primer will anneal securely to the plasmid DNA at higher temperatures. The reaction is stopped after 30 cycles rather than 40 to minimize the chance of errors being introduced by nucleotide mis-incorporation and non-specific annealing.

PCR method

(1) Turn on thermal cycler power.

(2) (a) Back-up template preparation:

Boil enough 1:500 pTZLT18 template plasmid DNA for several reactions for 5 min at 100°C in the thermal cycler. When using the same template for several reactions, the DNA can be heated in bulk so losses through condensation are minimized. The DNA is heated to inactivate any nucleases that may be present in the sample. Alternatively, the back-up template plasmid DNA can be treated the same as in 2(b).

(b) Preparation of DNA template stock from Experiment 2:

For previously isolated and purified template plasmid from Experiment 2, mix in a sterile 200 ml dome-top PCR tube,

19 µl autoclaved distilled water (nuclease-free water)

10 µl (~4 ng) plasmid DNA template (or an amount to total 40 ng DNA/29 µl)

Heat at 100 °C in the thermal cycler for 5 m.

(3) Remove the tube from the thermal cycler and centrifuge it for 30 s to collect the condensate accumulated on the tube lid.

(4) (a) Back-up reaction:

Label a 200 µl PCR reaction tubes 'B-u'. Distribute 29 µl boiled pTZLT18 DNA to each tube. Proceed to add remaining reagents as listed below.

(b) DNA template plasmid isolated in Experiment 2:

add reagents listed below to the 29 µl boiled DNA in the reaction tube. Make sure the tube is labeled with appropriate team initials.

For PCR mix (final total volume will be 50 µl)

To 29 µl boiled template in a dome-top 200 µl tube add:

5 µl LTB-F(NcoI) primer (4 µM stock, final concentration 0.4 µM)

5 μl LTB-R BamHIprimer (4 μM stock, final concentration 0.4 μM)

5 μl 10× *Pfx* PCR buffer (final concentration 1×)

3 μl 5 mM dNTP mix (containing 5 mM dATP, 5 mM dCTP, 5 mM dGTP, and 5 mM dTTP in 1× PCR buffer* (final concentration 0.3 mM)

2 μl 25 mM $MgSO_4$† (final concentration 1 mM).

Add nuclease-free water to make up to 49 μl if necessary. Note the amount added. The final μl will be *Pfx* DNA Polymerase added in step (7) below.

(5) Mix well by a quick vortex or pipetting up and down.

(6) Negative controls:

two groups should each make the controls substituting autoclaved distilled water for the template plasmid DNA. A positive control using 'bystander' DNA (for example, a plasmid of completely different origin but similar size and DNA distribution) in place of pTZLT18 DNA can also be set up.

(7) Keeping the reaction mix on ice, add 1 μl *Pfx* DNA polymerase (or 2 μl polymerase diluted 1:1 with 1× PCR buffer) into the reaction tube.

(8) Cap the tubes, check the labels, centrifuge the tubes 30 s to collect the contents, then place them on ice. If flat-top PCR tubes have been used, it is necessary to overlay the reaction mixture with 20 μl sterile mineral oil to prevent losses due to condensation.

(9) Set the thermal cycler to 85°C. When all tubes are ready, dry each tube with a KimWipe and place them in the thermal cycler.

(10) Position the 'hot start' or 'quick start' lid and begin the following protocol:

Pfx Protocol for LTB	*Explanation of step*
(step 1) 94°C, 2 m	main denaturation
(step 2) 94°C, 45 s	cycle denaturation
(step 3) 40°C, 45 s	cycle annealing
(step 4) 68°C, 45 s	cycle extension
(step 5) 29× to (2)	30 cycles (steps 2, 3 and 4)
(step 6) 72°C, 5 m	finishing (to ensure full-length product)
(step 7) 15°C, hold	keep cold
(step 8) END	

---

*Use 1.5 μl 10 mM dNTP mix.

†Use 1 μl 50 mM $MgSO_4$. The final concentration must be as indicated.

(11) After the reaction sequence has been completed, remove the tubes and place the samples in a storage box in the freezer. This sequence can remain on hold at 15°C overnight if necessary.

Preparation for experiment

(1) DNA standards: a 1 mg/ml (1 µg/µl) DNA stock such as λ DNA or purified calf thymus DNA can be used. Add x µl stock to 1 ml 10 mM Tris–HCl, pH 8.5 or nuclease-free water to obtain an x µg/ml (x ng/µl) standard.

(2) Ethidium bromide test solution: 1 ng ethidium bromide/µl nuclease-free water or 10 mM Tris–HCl, pH 8.5, or this can be prepared by dilution from a 10 mg/ml ethidium stock solution in water. Ethidium bromide is not very soluble so it is very important to vortex the solutions frequently, especially before measuring.

(3) Template DNA: pTZLT18 and/or sample from Experiment 2, *ca* 0.4 ng/µl. Dilute with nuclease-free water from a known concentration of DNA obtained from a midiprep. Ten mM Tris–HCl may also be used. Do not use EDTA as it will chelate the Mg^{2+} and inhibit the reaction.

(4) Primers: 4 µM LTB-F (NcoI) and 4 µM LTB-R (BamHI). Resuspend commercially synthesized primers to 100 µM then dilute to a 10 µM stock. The 10 µM stock can be diluted to 4 µM for use, or directly dilute the 100 µM stock 1:25 (v/v). Primers can be obtained from several suppliers, e.g. Integrated DNA Technologies (http://www.idtdna.com). For instructions for making Tris buffers see Experiment 2.

References

Ausubel, F. M., Brent, R., Kingston, R. E., Moore, D. D., Seidman, J. G., Smith, J. A. and Struhl, K. (1999). *Short Protocols in Molecular Biology*. New York: John Wiley.

Brock, T. D. (1994). *Life at High Temperatures*, p. 31. Yellowstone National Park, Wyoming: Yellowstone Association for Natural Science, Hinsory & Education, Inc.

Mullis, K. (1990). The unusual origin of the polymerase chain reaction. In *Scientific American* April, pp. 56–65..

Mullis, K., Faloona, F., Scharf, S., Saiki, R., Horn, G. and Erlich, H. (1986). Specific enzymatic amplification of DNA in vitro: the polymerase chain reaction. *Cold Spring Harbor Symposia on Quantitative Biology* LI, 263–273.

Sambrook, J. and Russell, D.W. (2000). *Molecular Cloning: a Laboratory Manual*. Cold Spring Harbor, NY: Cold Spring Harbor Laboratory Press.

EXPERIMENT 4

(a) Agarose Gel Electrophoresis
(b) Recovery of PCR Product

Introduction

Agarose gel electrophoresis

Agarose gel electrophoresis (AGE) is the most common technique used to separate fragments of DNA that differ in size. Agarose is a purified linear galactan, a galactose polymer isolated from agar or recovered directly from agar-bearing marine algae. Agar is a mixture of polysaccharides extracted from certain red seaweeds. During AGE, particles or fragments move with an applied electric current through an agarose gel slab (matrix), which acts as a molecular sieve. Smaller molecules pass through the matrix more rapidly than large molecules. Since all DNA has about the same molecular charge/mass ratio ($1 PO_4^{2-}$ per nucleotide), separation of DNA fragments is based entirely on size differences. There is a linear relationship between the rate of migration of a linear DNA fragment of a given size and the concentration of agarose, based on the log of the electrophoretic mobility of the DNA and the gel concentration (see Sambrook and Russell, 2000, Vol 1, pp. 5.5–5.7 for a detailed discussion). At low voltages, the rate of migration of linear DNA fragments is proportional to the voltage applied. However, at higher voltages, the rate of mobility increases at different rates. The effective voltage range in an agarose gel is 5–8 V/cm (no more than 85 V/10 cm gel).

The concentration of agarose can be varied to change the size of the gel matrix (the fibrous mesh) to provide good separation in the size range of interest. This is shown in

Table 4.1 Suggested agarose concentration for separation of DNA fragments

Size range (base pairs)	Agarose concentration (% w/v)
1000–23 000 bp	0.6%
800–10 000 bp	0.8%
400–8000 bp	1.0%
300–7000 bp	1.2%
200–4000 bp	1.5%
100–3000 bp	2.0%

Table 4.1. Note that increasing the concentration of agarose 'tightens' the mesh, which results in improved separation of smaller fragments.

The conformation of the DNA, which can be superhelical, nicked or linear, also influences the rate of migration (Parker, Watson and Vinograd, 1977; see Experiment 6 for a further discussion of this point).

Visualization

DNA can be visualized in the gel with ethidium bromide. It intercalates (is entrapped) into the grooves of double-stranded DNA and fluoresces in UV light. The high concentration of DNA in a band on a gel will fluoresce brightly against a background of diffuse ethidium bromide in the gel matrix.

Warning: because ethidium bromide intercalates into the grooves of DNA it is a potential mutagen (mismatches may occur in the stretched helix). *Wear gloves when handling gels containing ethidium bromide.* To dispose of the gel wrap it in plastic cling film and place it in a securely closed plastic bag with a paper towel. It can then be disposed of with the regular waste.

Warning: ultraviolet light is a mutagen because ionizing radiation causes cross-links in both DNA and proteins and will burn skin and retinas. *Wear goggles and use a shield when viewing gels.*

The total size of the PCR product itself should be approximately 350–400 bp, representing only the LTB portion of the operon with cleavage sites added. The percentage of agarose in the gel has been adjusted so that the PCR fragment runs closer to the middle of the gel. After separating the PCR fragment by electrophoresis and verifying its size and purity, the fragment will be recovered from the gel. Any non-specific transcripts present in the reaction mix should be visible as extra, lower-concentration bands on the gel. By excising only bands of the correct molecular weight the fragment is purified away from non-specific DNA fragments of different size. Template DNA will not be visible on the gel, since its original concentration was very small and the fragment has been amplified enormously. Any dNTPs and primers remaining in the reaction mix will have run much faster than the fragment and most likely will have run off the end of the gel. In addition, DNA polymerase should be well separated from the PCR product. This is particularly important because the polymerase can fill in the recessed 3' termini created by subsequent digestion of the product with restriction endonucleases (Experiment 5), making ligation into the cloning vector problematic. Electrophoretic isolation, excision of the fragment band, and a kit purification should provide a good cloning insert.

Recovery

DNA can be recovered from agarose gels by a variety of techniques, depending on convenience.

(1) It can be electroeluted, a technique in which a slice of agarose containing the DNA band is placed in a dialysis bag inside an electrophoresis apparatus. As the current passes through the gel the DNA sample migrates out of the gel and is trapped inside the dialysis bag in the buffer.

(2) After the DNA band is sliced from the gel, the agarose can be melted in sodium iodide buffer, then bound to a membrane or to a resin. The melted agarose is rinsed away and the DNA is eluted from the membrane or resin.

(3) A specially formulated low-melting agarose can be used for the electrophoresis. When warmed to 60°C, the agarose liquefies. The enzyme agarase is added to digest the agarose polymer, leaving small fragments. DNA can be precipitated out of the resulting suspension.

(4) A 0.45 μm centrifugation-filtration tube can be used to separate the DNA (which passes through the filter during centrifugation). The agarose is trapped on top of the filter. This technique is well-suited to fragments from 500 to 5000 bp.

Materials and Methods

Materials

AGE

♦ P-20 and P-200 Pipetmen, sterile white or yellow tips

♦ Horizontal AGE apparatus (two teams per eight-well gel)

♦ Gel casting platform, gel combs (slot formers), power supply

- Agarose, ultra pure

- 125 ml Erlenmeyer (conical) flask

- 50× Tris–acetate–EDTA (TAE) or 10× Tris–borate–EDTA (TBE) electrophoresis buffer diluted to 1× (final concentration of 1× TAE is 40 mM Tris acetate/1 mM EDTA/pH 8.0)

- 6× or 10× AGE loading buffer (AGE-LB)

- Ethidium bromide stock (10 mg/ml TE buffer)

- PCR product from Experiment 3

- PCR markers, 25 µl + 5 µl 6× AGE loading buffer

Recovery of PCR product

- UV transilluminator

- Microcentrifuge

- Clean single-edge razor blades

- Goggles

- Latex gloves

- Plastic cling film

- Number 30 sunscreen or a face visor/facemask

- QIAquick Gel Extraction kit (QIAgen catalog no. 28704)

- 1.5 ml microcentrifuge tubes

If a kit is not used:

- Spin-Vac or desiccator

- UV transilluminator

- Microcentrifuge

- Clean single-edge razor blades

- 1.5 ml microcentrifuge tubes

- Glass rod or small tissue grinder for microcentrifuge tube

- Goggles

◆ Latex gloves

◆ Number 30 sunscreen or a face visor/facemask

◆ 0.45 μm filter units

◆ TE buffer (10 mM Tris–HCl, pH 8.5/1 mM EDTA)

◆ TE-saturated *n*-butanol

◆ 100% ethanol, stored in freezer

◆ 70% ethanol, stored in freezer

Method

AGE

(1) Make 1× TAE buffer (final concentration 40 mM Tris Acetate/1 mM EDTA/pH8.0) by diluting 20 ml of 50× stock TAE to 1000 ml, or make a suitable dilution from any other concentration stock. Each gel will require 300 ml 1× buffer.

(2) For the 350–400 bp insert, prepare a 1.3% gel (see Table 4.1). In a 125 ml Erlenmeyer (conical) flask, mix 30 ml 1× TAE buffer with 0.4 g agarose. Melt the mixture in a microwave oven for 1–2 min until it boils, or boil it on a hot plate until the solution is clear. Remove the flask from the heat using an oven mitt. Swirl the mixture to mix it and let it cool to 60°C for safe handling. When the mixture is cool enough to handle, add 2–3 μl ethidium bromide stock (for a final concentration of 0.66 μg/ ml). The ethidium bromide is for visualization of DNA. Swirl to mix.

(3) Set up the gel casting platform and position the seals or dams according to the manufacturer's instructions. Pour the gel slowly and evenly into the center of the platform. Place the comb (well-former) in its slot about 2 cm from one edge, using the eight-well side.

(4) After the gel has hardened and become opaque (15–30 min), remove the seals or dams from the platform and remove the comb. Place the platform into an electrophoresis tank so the wells are closest to the negative electrode. Pour in 250 ml electrophoresis buffer, to cover the gel 1–2 mm above its surface.

(5) Remove the PCR reaction tubes from the freezer. Add 10 μl 6× AGE-LB to 50 μl reaction mixture. The mixture should be vortexed and can be heated to 70°C for

2 min to prevent aggregation. Make sure the tube is clearly marked. Do the same for the two control tubes.

(6) Load PCR marker into the first and last wells, 5–10 μl/well when using the eight-well gel. Alternatively, load marker into well no. 4 or 5.

(7) Load 20 μl of sample into each well. Note the positions of the wells containing the sample. A 60 μl sample will fill three wells.

(8) Load one control sample on each gel, making sure to note the position of the well.

(9) Attach the lid and leads so the DNA migrates to the anode (red, positive electrode) and electrophorese at 80–90 V (8 V/cm of gel) until the blue dye is centered and the orange is 1–2 cm from the bottom, about 30–40 min. The 400 bp fragment should appear just above the yellow dye band between the blue and yellow bands.

(10) Shut off the power supply then pull the leads. Remove the lid and, wearing gloves, remove the gel and place it on a sheet of plastic wrap. Gel may be destained briefly in distilled water to remove excess background before viewing, although this is not necessary.

(11) Place gel, on plastic film, on the glass plate of an ultraviolet transilluminator. *Use goggles and a face mask or sunscreen. Cover viewer with UV-proof glass before turning on the light.* Turn on UV, *work quickly here* – too much UV light will cause DNA sample breakage and burn unprotected face or hands.

(12) Record and label the results. Label the wells with your initials and the sample name. A photograph will also be taken and copies distributed for your notebook.

Preferred method: recovery of PCR product

Instructions* are for a QIAgen Gel Extraction Kit, but Promega Wizard Gel Extraction Kit or other brand may be used, following the specific kit instructions.

(1) Weigh a colorless 1.5 ml tube. Mark the weight on the side of the tube.

(2) Clean a razor blade with ethanol, wipe it dry and carefully cut the PCR product from the gel. Trim away any excess agarose. *Use goggles, gloves, and no. 30 or higher UVA–UVB sunscreen on face or use a facemask* to avoid getting burned while leaning over the box to cut out the band.

---

*Reproduced by permission of QIAgen.

(3) Slide the excised gel slice in the tube and re-weigh. Calculate the weight (volume) of the gel slice.

(4) Add 3 vols QIAgen kit buffer QG to 1 vol of gel. For example, for 0.1 g (100 mg) gel add 0.3 ml (300 μl) buffer. The maximum gel slice for one column is 400 mg (0.4 g).

(5) Incubate at 50°C for 10 min or until the gel slice has completely dissolved. To help the gel dissolve, mix by vortexing every 2–3 min. The mixture should be yellow. If the color of the mixture is orange or violet the pH is not correct. Add 10 μl of 3 M sodium acetate pH 5.0 and mix. The color should return to yellow.

(6) Add one gel volume of isopropanol to the liquefied sample and mix. For example, if the slice is 100 mg add 100 μl isopropanol. This step increases the yield of DNA fragments smaller than 500 bp and larger than 4 kb. We expect a 350–400 bp fragment, so the use of isopropanol is optional, but does seem to help the yield.

(7) Place a QIAquick spin column in a 2 ml collection tube.

(8) Apply the solubilized gel sample to the QIAquick column and centrifuge for 1 min. The DNA will bind to the membrane while agarose and small molecules including ethidium bromide, will elute into the collection tube. Further washing is unnecessary for the reactions undertaken in the next few experiments.

(9) Decant and discard the flow-through.

(10) Place the QIAquick column back in the same collection tube. Add 0.75 ml (750 μl) QIAgen kit buffer PE to the column and centrifuge for 1 min.

(11) Decant and discard the flow-through. Replace the collection tube and centrifuge for an additional 1 min. The eluent must be removed from the collection tube to prevent any ethanol from vaporizing back into the membrane. The additional centrifugation will remove residual ethanol that was present in buffer PE. Ethanol would inhibit the restriction endonuclease reaction in Experiment 5.

(12) Place the spin column into a clean, sterile 1.5 ml microcentrifuge tube.

(13) Add 50 μl QIAgen kit buffer EB (10 mM Tris–HCl, pH 8.5) or 50 μl H_2O to the center of the membrane and centrifuge the column for 1 min (for increased DNA concentration add 30 μl EB to the center of the membrane and let it stand for 1 min before centrifuging). No EDTA should be used. It will inhibit subsequent enzymatic reactions.

Non-kit preparation of PCR product

(1) Wipe off a razor blade with ethanol and carefully cut your PCR product from the gel. Trim away any excess agarose. *Use goggles and no. 30 or higher UVA–UVB sunscreen on your face and hands* so you do not get burned while leaning over the box to cut out the band.

(2) Add 1–2 vols TE buffer.

(3) Sterilize a glass rod in ethanol. Use it to macerate the gel.

(4) Transfer the gel and buffer to a 0.45 μm filter unit.

(5) Centrifuge for 5 min.

(6) Add 1 vol buffer to the filter

(7) Centrifuge again.

The DNA and accompanying ethidium bromide will be in the eluent. The ethidium bromide must be removed prior to freezing. This can be accomplished by adding ice-cold 95% ethanol to the eluent then freezing to precipitate the DNA. The mixture is then centrifuged, washed and re-pelleted several times with 70% ethanol. The pellet should be dried well using a Spin-Vac or vacuum desiccator, and resuspended in a small amount Tris buffer (10 mM Tris, pH 8.5) or sterile distilled water. No EDTA should be used as it could inhibit the restriction endonuclease reaction.

Preparation for experiment

(1) 50× TAE electrophoresis buffer, pH 8.5 (500 ml): 121 g Tris base; 28.6 ml glacial acetic acid; 18.6 g Na₂EDTA·2H₂O; H₂O to 500 ml; 1× working solution is 40 mM Tris-acetate/2 mM EDTA. TBE may be used instead of TAE. Migration patterns differ in different buffers. 10× TBE electrophoresis buffer: 108 g Tris base; 55 g boric acid; 40 ml 0.5 M EDTA pH 8.5; H₂O to 1 l. 1× working solution is 89 mM Tris/89 mM boric acid/2 mM EDTA.

(2) 6× AGE gel loading buffer: a variety of gel loading buffers are commercially available. The Promega 6× AGE loading buffer catalog no. G1881 works very well, is premixed and contains orange G, bromophenol blue and xylene cyanol.

6× Buffers (Sambrook and Russell, 2000 vol. 3, p. A1.19):

(a) Type I – 0.25% (w/v) bromophenol blue, 0.25% (w/v) xylene cyanol FF, 40% (w/v) sucrose in water;

(b) Type II – Substitute 15% (w/v) Ficoll (type 400; Pharmacia) in water for sucrose in water;

(c) Type III – Substitute 30% (w/v) glycerol in water for sucrose in water;

(d) Type IV – Eliminate the xylene cyanol FF from the type I recipe.

(3) It is useful to aliquot individual portions with a few (2–10) μl extra for each solution used from the kit to avoid contamination of the bulk solution and simplify distribution.

References

Parker, R. C., Watson, R. M. and Vinograd, J. (1977). Mapping of closed circular DNAs by cleavage with restriction endonucleases and calibration by agarose gel electrophoresis. *Proc. Natl Acad. Sci.* 74, 851–855.

Sambrook, J. and Russell, D.W. (2000). *Molecular Cloning: a Laboratory Manual*. Cold Spring Harbor, NY: Cold Spring Harbor Laboratory Press.

EXPERIMENT 5

(a) Restriction Digest of LTB Insert and Vector
(b) Plate Preparation

Introduction

The LTB gene that has been amplified by PCR must be placed into a suitable vector (carrier) then moved into a bacterial host. The bacterial host cell must be capable of propagating the vector, transcribing the gene, and translating the transcription message into the properly folded, functional protein. These steps in the host cell are termed 'gene expression'. In order to carry out the transfer of the gene into the host cell, the vector, a circular plasmid, must be cut and the newly created ends must be joined to the ends of the gene insert (PCR product). This joining is accomplished by cleavage of the DNA at specific sites which can then be re-united with each other or, in this case, spliced with a DNA insert. The splicing is based on specific base-pairing that positions the insert on the cut ends of the vector plasmid, then ligation of the insert to the plasmid. The insertion of 'foreign' DNA into a DNA sequence is termed 'recombination' and this step is therefore the basis of recombinant DNA technology.

Cutting DNA at particular nucleotide sequences is accomplished by restriction endonucleases that recognize and cleave double-stranded DNA (dsDNA) at specific sequences, usually palindromes of 4–8 bp. The enzymes are named by their original bacterial source, for example, *Eco*R1 from *E. coli*, *Bam*H$_I$ from *Bacillus amyloliquefaciens* H, *Hae*III from *Haemophilus aegyptus*, or *Hind*III from *Haemophilus influenzae* d. The cleavage may leave blunt ends (both strands cut at the same position) or 'sticky' ends, in which there is a 5′ overhang in which unpaired bases protrude on the 5′ ends of the cut DNA.

Examples

*Bam*HI (from *Bacillus amyloliquifaciens*) recognizes 5′...GGATCC...3′ and cleaves 5...G|GATCC...3′ to leave a 5′ overhang of 5′ GATCC 3′ on each strand,
    3′...CCTAG|G...5′.

*Nco*I (from *Nocardia corallina*) recognizes 5′...CCATGG...3′ and cleaves 5′...CC|ATGG...3′ }
3′...GGTA|CC...5′ } to leave a 5′ overhang of 5′ ATGG 3′ on each strand,

*Nde*I (from *Neisseria dendrificans*) recognizes 5′ CATATG...3′ and cleaves 5′...CA|TATG...3′ to leave a 5′ overhang of 5′ TATG 3′ on each strand,
    3′...GTAT|AC.

Note that in all cases the recognition sites are palindromic, meaning they read the same 5′→3′ on each strand. Put another way, they read the same forwards and in reverse. Restriction endonucleases are used by bacteria to cleave infective foreign DNA. They have been exploited to great advantage by genetic engineers to cut and splice inserts into vectors. If the vector (a plasmid that will carry the insert) is cut with the same endonuclease as the insert, the sticky ends of the insert complement the sticky ends of the vector. The insert can be dropped into the gap between the two sticky ends on the vector, shown in Figure 5.1. The plasmid to be used as a vector, pET 28b, has been designed and built specifically for cloning and expressing inserted genes. It carries an N-terminal His−Tag/T7−Tag site with a thrombin cleavage site for removing the tag from the expressed protein. It also carries an optional C-terminal His−Tag sequence. The plasmid is numbered by the pBR322 convention so the T7 expression region is reversed on the circular map. PBR322 is the plasmid from which the pET series of plasmids derives. It carries the kan gene (Kanamycin resistance). The plasmid vector pET28b has been chosen for this project because of its versatility.

Most restriction enzymes used in genetic engineering are endonucleases, requiring two to six base pairs beyond the recognition sequence for optimal cleavage. The additional basepairs stabilize the binding site, serving as a 'perch' for the enzyme as it binds and surrounds its recognition site on the DNA. In addition, three to six base pairs may be trimmed away during cleavage, and these base pairs should not be part of the target gene.

The optimal conditions vary with the endonuclease and have been worked out by commercial vendors. One must also consider how to remove or inactivate the enzyme after it has cleaved the DNA sample. Other issues include ability of the endonuclease to cleave methylated DNA (a trick used by the bacteria of origin to protect their own DNA)

pET-28a-c(+) Vectors

	Cat. No.
pET-28a DNA	69864-3
pET-28b DNA	69865-3
pET-28c DNA	69866-3

The pET-28a-c(+) vectors carry an N-terminal His•Tag®/thrombin/T7•Tag® configuration plus an optional C-terminal His•Tag sequence. Unique sites are shown on the circle map. Note that the sequence is numbered by the pBR322 convention, so the T7 expression region is reversed on the circular map. The cloning/expression region of the coding strand transcribed by T7 RNA polymerase is shown below. The f1 origin is oriented so that infection with helper phage will produce virions containing single-stranded DNA that corresponds to the coding strand. Therefore, single-stranded sequencing should be performed using the T7 terminator primer (Cat. No. 69337-3).

pET-28a(+) sequence landmarks

T7 promoter	370-386
T7 transcription start	369
His•Tag coding sequence	270-287
T7•Tag coding sequence	207-239
Multiple cloning sites	
(*Bam*H I - *Xho* I)	158-203
His•Tag coding sequence	140-157
T7 terminator	26-72
lacI coding sequence	773-1852
pBR322 origin	3286
Kan coding sequence	3995-4807
f1 origin	4903-5358

The maps for pET-28b(+) and pET-28c(+) are the same as pET-28a(+) (shown) with the following exceptions: pET-28b(+) is a 5368bp plasmid; subtract 1bp from each site beyond *Bam*H I at 198. pET-28c(+) is a 5367bp plasmid; subtract 2bp from each site beyond *Bam*H I at 198.

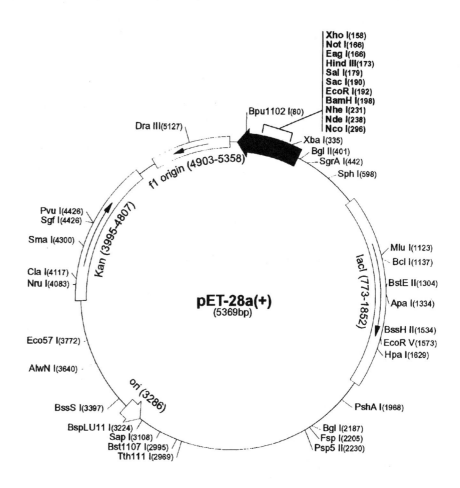

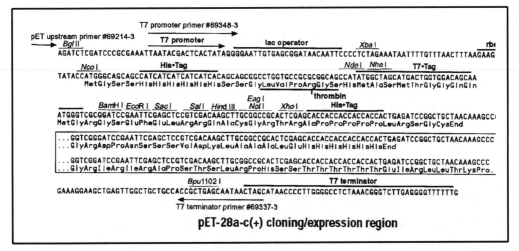

pET-28a-c(+) cloning/expression region

Figure 5.1 Plasmid vector set pET28a–c; pET28b has 5368 bp (reproduced by permission of Novagen Inc.)

and whether or not the endonuclease will cleave DNA sequences different from the desired sequence. The recognition and cut sites can be found in textbook tables and on the Internet at commercial supplier sites such as New England BioLabs, Promega, Calbiochem, etc. The enzymes generally require different reaction conditions (salt concentration, ionic strength, specific counter ions), but in many cases one enzyme is less fastidious than the other and can cut DNA reasonably well in several different reaction mixtures. The control, single-cut plasmid should be done with the enzyme acting in its non-optimal buffer to be sure it cleaves properly. It is also necessary to allow enough space (base pairs) between sites so the two enzymes used in a double digest do not sterically hinder each other. For example, the 98 bp spacing between the BamHI and NcoI sites is sufficient for a simultaneous double digestion. The 40 bp between the BamHI and NdeI sites requires sequential (one-at-a-time) digests.

Materials and Methods

Materials

Double digestion

- Thermal cycler, water bath or heat block set at 37°C

- P-20 Pipetmen and sterile white or yellow tips

- *Bam*HI, 20 U/μl

- *Nco*I, 20 U/μl

- New England BioLabs 10× BamHI digestion buffer or Promega 10× buffer D (see Preparation for Experiment section)

- 100X acetylated bovine serum albumin (BSA; supplied with enzyme) or 50× diluted in 1× digestion buffer from 100× BSA

- 200 μl sterile PCR tubes, preferably dome-cap

- nuclease-free distilled water

- Purified PCR product from previous experiment (~1–2 μg) per team

- PET28b vector (~2–4 μg)

- 6× AGE-LB

Plate preparation

♦ Autoclave

♦ Sterile Petri dishes

♦ Bactotryptone (pancreatic digest of casein, rich in Trp)

♦ Yeast extract

♦ NaCl

♦ Agar
  or prepared Miller Luria–Bertani (LB) agar or Miller LB broth with added agar

♦ Kanamycin stock (30 mg/ml dH$_2$O)

Method

Double digestion of PCR fragment insert (I)

Insert (I) is the PCR fragment having a nucleotide sequence corresponding to the LTB gene. The insert reaction should be prepared simultaneously with the vector reaction so they can be incubated at the same time. The vector plasmid is pET28b. You could prepare a 'master mix' containing sufficient digestion buffer, enzymes and BSA for several reactions. In that case, use 9–10 μl master mix per tube and add sample with sufficient water to make a 50 μl total reaction mixture.

It may be useful to do a preliminary quick ethidium bromide spot test to determine the approximate concentration of PCR product DNA.

(1) Mix in a 200 μl PCR tube
   5 μl 10× digestion buffer
   1.5–2 μl BamHI (10 U/μg insert DNA)
   1.5–2 μl NcoI (10 U/μg insert DNA)
   1 μl 50× acetylated BSA or 0.5 μl 100× (to bind any contaminants that might inhibit the endonucleases)
   x μl purified PCR product, thawed (enough to provide about 2 μg DNA)
   y μl nuclease-free water (to make reaction mixture up to 50 μl total).
   Up to 40 μl of gel-purified PCR product can be used for this reaction. For $x = 40$ μl PCR product, $y = 0$ μl water. For smaller amounts PCR product, add water to make the total reaction mixture 50 μl.

(2) Incubate in the thermal cycler or waterbath at 37°C for 1.5–2 h

(3)   After incubation, heat to 65°C for 5 min to inactivate the enzyme protein.

(4)   Add 10 µl of 6× AGE-LB to the 50 µl reaction mix and freeze immediately.

Double digestion of pET 28b vector (V)

(1)   Mix in a 200 µl PCR tube
        5 µl 10× buffer
        1.5–2 ml BamHI (10 U/µg vector DNA)
        1.5–2 µl NcoI (10 U/µg vector DNA)
        1 µl 50× BSA; *x* µl pET28b plasmid (enough to provide about 3 µg vector DNA)
        y µl nuclease-free water (to make reaction mixture up to 50 µl total).
      Thirty to forty microliters of crude vector midiprep can provide sufficient digested
      vector for several ligations and transformations.

(2)   Incubate in the thermal cycler at 37°C for 1.5–2 h

(3)   After incubation, heat to 65°C for 5 min to inactivate the protein enzymes.

(4)   Add 10 µl of 6× AGE-LB to the 50 µl reaction mix and freeze immediately.

Controls

Control digestions can be run simultaneously. Controls should include: (C1) vector DNA
digested with *NcoI* only; (C2) vector DNA digested with *Bam*HI only; (C3) vector DNA
mixed with nuclease-free water substituted for enzymes (optional).

Controls can be run using 5–10 µl pET28b in a 25 µl total reaction mixture. At the end
of the reaction add 5 µl of 6× AGE-LB to the 25 µl reaction mix and freeze.

LB-Kan30 plate preparation

Following restriction endonuclease digestion and ligation of the insert into the vector,
the recombinant plasmid must be transported into a bacterial host, a procedure called
'transformation'. Bacteria containing recombinant plasmid will form colonies on the
selective agar. This process requires agar plates to be made for use in subsequent
experiments. Pre-packaged LB (Miller) agar and broth are also available, and can be used
in place of individual ingredients.

Method

LB agar: 5 g Bacto tryptone (pancreatic digest of casein, rich in Trp); 2.5 g yeast extract; 5 g NaCl; and 7.5 g agar.

(1) Add solid ingredients to a 2 l flask, add 500 ml distilled water and swirl to disperse solids. A 500 ml batch of agar is sufficient to fill 1 sleeve of 20 sterile Petri dishes (plates).

(2) Cover flask with aluminum foil and a piece of autoclave tape. Autoclave for 20–30 min using the liquids cycle.

(3) Remove the flask and place in a room temperature water bath. Cool to 60°C (cool enough to handle but not yet solidified).

(4) Add 0.5 ml Kanamycin (30 mg/ml stock) for a final concentration of 30 μg/ml media. Swirl to mix. Avoid bubbles.

(5) Remove the plates from the sterile plastic sleeve and line them up *closed* on the bench top. Save the sleeve. Grasp the flask with one hand, at the neck, open it, flame the mouth of the flask and, with your other hand, open one plate at a time. Quickly pour the agar mixture to about a third to a half full. Cover the plate. The lid may be set slightly askew to allow water vapor to escape. To maintain sterility, flame the top of the flask before pouring each plate. The agar surface may be flamed quickly with the Bunsen burner after pouring to remove any bubbles.

(6) Allow the agar to solidify for about 15 min. It will become opaque. When the agar has solidified, invert the plates so the condensate does not drip on the agar surface. Leave the inverted plates on the bench top for 24 h to cure.

(7) Label the plates at the edge of the bottom, noting 'LB-Kan-30' and the date poured. Stack them carefully into the saved sleeve, tape it shut and store the plates in the refrigerator.

Preparation for experiment

(1) Digestion buffer is normally supplied with the enzymes. Buffers chosen should work for both enzymes. If New England BioLabs (Beverly, MA, USA) enzyme is used, use the *Bam*HI 10× buffer supplied, containing 1500 mM NaCl/100 mM Tris-HCl/100 mM $MgCl_2$/10 mM DTT (dithiothreitol). If Promega (Madison, WI, USA) enzyme is used, the 10× buffer D, supplied with *Nco*I, works for both enzymes. It

contains 1500 mM NaCl/60 mM Tris–HCl/60 mM MgCl$_2$/10 mM DTT (dithiothreitol).

(2) Prepared Miller LB agar can be used for the preparation section. Add the requisite amount of water; autoclave to dissolve and sterilize it. Add 0.5 ml of 30 mg/ml Kanamycin stock solution to 500 ml slightly cooled agar. Swirl to mix. Miller LB broth with added agar can be used and treated as above.

(3) Kanamycin stock (30 mg/ml dH$_2$O): 300 mg (0.3 g) Kanamycin (Sigma no. K4000 is satisfactory); 10 ml sterile distilled water. Make 0.5 ml aliquots in clean 1.5 ml microcentrifuge tubes. Label as '30 mg/ml Kan, 0.5 ml'. Keep frozen until required.

EXPERIMENT 6

(a) Purification of Digested Insert and Vector by Agarose Gel Electrophoresis
(b) Recovery of Digested Insert and Vector

Introduction

Purification

Both LTB gene insert and pET28b vector will be gel purified. Agarose gel electrophoresis is the most efficient way to simultaneously check the extent of digestion, verify DNA sizes and purify the digested DNA away from enzymes and other proteins in the reaction mixture.

DNA can be found in three forms: form I, supercoiled (superhelical); form II, predominantly nicked circular (open circles), and form III, mainly linear. Supercoiled preparations such as plasmids generally contain nicked circular forms that occur when the supercoil is unwound by a single-stranded nick in the DNA. The open circle is seen as a second band on the gel. The supercoiled plasmid can be nicked in a controlled manner in the presence of low concentrations of ethidium bromide (which stretches the DNA open where it intercalates) and a low concentration of restriction endonuclease (Barzilai, 1973; Osterlund *et al.*, 1982). Each of these forms has a different mobility on a gel, the mobility depending on the concentration of agarose, the applied voltage, the presence of ethidium bromide in the gel, the molecular size and conformation of the DNA and the composition of the electrophoretic buffer. Migration of a linear fragment of a given size in various concentrations of agarose was discussed in Experiment 3.

Conformation (form I, II or III) modifies mobility because each form has a different cross-section, and thus a different frictional coefficient when being pulled through the gel, even though the molecular weight of the fragment is the same (Thorne, 1967). In addition, the buffer composition may alter intra-chain interactions, thereby altering mobility of each form in different buffers. In general, the supercoiled form I will migrate fastest because of its compact form. Whether form II (open circle) migrates faster or slower than form III (linear) is very dependent on the buffer ionic strength and composition. Ethidium bromide, because it stretches the helices of the DNA, tends to increase the effective cross-section of the molecule, thus retarding mobility.

Digesting the vector plasmid 'linearizes' it. The pET28b plasmid vector (V), 5368 bp, will be linear, no longer supercoiled, and will, in addition, have lost 98 bp due to excision of a fragment between the *Bam*HI and the *Nco*I sites. It will run slightly faster than single digested vector. Controls in which water has been substituted for enzyme should remain supercoiled. These preparations are least mobile and characteristically have two bands, representing form I (supercoil) and form II (nicked circular).

The LTB gene insert (I) should have become slightly shorter, losing only 3–4 bp from each end, but will not be otherwise changed in form, having been linear to begin with.

Recovery of insert and vector

The recovery and concentration of insert DNA and pET28b vector DNA will be accomplished by gel extraction using a gel extraction kit.

Materials and methods

This experiment is performed in the same way as Experiment 4. See Experiment 4 for a non-kit recovery protocol.

Materials

AGE

- UV transilluminator
- Goggles

- Latex gloves

- P-20 and P-200 Pipetmen, sterile white or yellow tips

- Horizontal gel electrophoresis apparatus (two teams per eight-well gel)

- Gel casting platform, gel combs (slot formers), power supply

- Number 30 sunscreen or face visor/facemask

- 125 ml Erlenmeyer (conical) flask

- Agarose

- 50× TAE or 10× TBE electrophoresis buffer diluted to 1× (final concentration of 1× TAE is 40 mM Tris acetate/1 mM EDTA/pH 8.0)

- 6× AGE loading buffer (6× AGE-LB)

- Ethidium bromide stock [10 mg/ml nuclease-free (autoclaved) water]

- Digested insert LTB gene (Idd) in AGE-LB from Experiment 5

- Digested vector pET28b (Vdd) in AGE-LB from Experiment 5

- Control C1 (Nco) in AGE-LB from Experiment 5

- Control C2 (Bam) in AGE-LB from Experiment 5

- Control C3 (pET28b, undigested)

- PCR markers, 25 μl+5 μl 6× AGE-LB

- 1 kb ladder, 25 μl+5 μl 6× AGE-LB

- Plastic cling film

Recovery of PCR product

- Microcentrifuge

- Clean single-edge razor blades

- Gel extraction kit (QIAgen QIAquick no. 28704 or no. 28104 w/buffer QG or Promega Wizard gel extraction kit)

- Sterile 1.5 ml microcentrifuge tubes

Method

AGE

(1) Make 1× TAE buffer (final concentration 40 mM Tris acetate/1 mM EDTA/pH 8.0) by diluting 20 ml of 50× stock TAE to 1000 ml, or make a suitable dilution from any other concentration stock. Each gel will require 300 ml 1× buffer.

(2) For the 350–400 bp insert (Idd), prepare a 1.3% gel (see Table 1). For the digested pET28b vector (Vdd), a separate 1.2% gel may be used. Alternatively both Idd and Vdd can be run on the same 1.2% gel for 30–40 min. For a 1.2% gel, use 0.36 g agarose in 30 ml 1× buffer. In a 125 ml Erlenmeyer (conical) flask, mix 30 ml 1× TAE buffer with 0.4 g agarose for a 1.3% gel. Melt the mixture in a microwave oven for 1–2 min until it boils, or boil it on a hot plate until the solution is clear. Remove the flask from the heat using an oven mitt. Swirl the mixture to mix it and let it cool to 60°C for safe handling. When the mixture is cool enough to handle, add 2–3 µl ethidium bromide stock (for a final concentration 0.66 µg/ml). Swirl to mix.

(3) Set up the gel casting platform and position the seals or dams according to the manufacturer's instructions. Pour the gel slowly and evenly into the center of the platform. Place the comb (well-former) in its slot about 2 cm from the end, using the eight-well side.

(4) After the gel has hardened and become opaque (15–30 min), carefully remove the seals or dams from the platform and remove the comb. Place the platform into an electrophoresis tank so the wells are closest to the negative electrode. Pour in 250 ml electrophoresis buffer, covering the gel 1–2 mm above the surface.

(5) Remove the digestion reaction mixtures from the freezer. The mixtures should be vortexed and can be heated to 70°C for 2 min to prevent aggregation. Make sure each tube is clearly marked. Do the same for the single digest control tubes.

(6) Load PCR marker into the first well, 5–10 µl/well when using the eight-well gel. Alternatively, load marker into well no. 4 or 5 to separate two groups' samples.

(7) Use a 1.3% gel for digested insert (Idd). Load 20 µl of Idd reaction mix into each of three adjacent wells, using the entire reaction mixture. Note the contents of each well.

(8) Load a 1.2% gel for doubly digested vector (Vdd) and vector single digestion control samples C1 and C2. A good pattern to compare single with double digested plasmid is: well 1 – 1 kb ladder, 6–8 µl; well 2 – control C1 (Nco), 20 µl; well 3 – Vdd, 20 µl; well 4 – Vdd, 20 µl; well 5 – Vdd, 20 µl; well 6 – control C2 (Bam), 20 µl; well 7 – control C3 (pET 28b undigested), 20 µl; well 8 – 1 kb ladder, 5–10 µl.

(9) Attach the lid and leads so the DNA migrates to the anode (red, positive electrode) and electrophorese at 80–90 V (8 V/cm of gel) until the blue dye is centered and the orange is 1–2 cm from the bottom, about 40–60 min for digested insert in a 1.3% gel and 60 min for digested vector in a 1.2% gel. The key is the positions of the dye bands. The 400 digested bp fragment should appear above the yellow-orange dye band and below the bromophenol blue band. Digested (linear) vector samples can be seen as single bands above the bromophenol blue dye, lying between the xylene cyanol dye band and the bromophenol blue dye band.

(10) Shut off the power supply then pull the leads. Remove the lid and, wearing gloves, remove the gel and place it on a sheet of plastic cling wrap. Gel may be destained briefly in distilled water to remove excess background before viewing, although this is not generally necessary.

(11) Place gel, on plastic cling film, on the glass plate of a ultraviolet (UV) transilluminator. *Use goggles and cover viewer with UV-proof glass before turning on the light, or use a face visor/facemask.* Turn on UV; *work quickly here* – too much UV light will cause DNA sample breakage and burn unprotected face or hands.

(12) Record the results. Label the wells with your initials and the sample name. A photograph will also be taken and copies distributed for placement in your notebook. See p. 54 for analysis of results.

(13) The gel should be wrapped in plastic cling film and placed it in a plastic bag with a paper towel for disposal in normal waste. The amount of ethidium bromide remaining in the gel is negligible since most of it has intercalated into the excised DNA.

Recovery of PCR product

Instructions are for a QIAgen Gel Extraction Kit*, but Promega Wizard or other brand may be used, following the specific kit instructions. Instructions for an alternate recovery method are given in Experiment 4.

(1) Weigh a sterile 1.5 ml tube. Mark the weight on the side of the tube.

(2) Clean a razor blade with ethanol, wipe dry and carefully cut the PCR product from the gel. Trim away any excess agarose. *Use goggles, gloves, and no. 30 or higher UVA–UVB sunscreen on face or use a face visor/facemask* to avoid getting burned while leaning over the box to cut out the band.

*Reproduced by permission of QIAgen.

(3) Slide the excised gel slice into the previously weighed tube and re-weigh. Calculate the weight (volume) of the gel slice.

(4) Add 3 vols QIAgen kit buffer QG to 1 vol of gel. For example, for 0.1 g (100 mg) gel add 0.3 ml (300 µl) buffer. The maximum gel slice for one column is 400 mg (0.4 g).

(5) Incubate at 50°C for 10 min or until the gel slice has completely dissolved. To help the gel dissolve, mix by vortexing every 2–3 min. The mixture should be yellow. If the color of the mixture is orange or violet the pH is not correct. Add 10 µl of 3 M sodium acetate pH 5.0 and mix. The color should return to yellow.

(6) Add 1 gel volume of isopropanol to the liquefied sample and mix. For example, if the slice is 100 mg add 100 µl isopropanol. This step increases the yield of DNA fragments smaller than 500 bp and larger than 4 kb. A 350–400 bp fragment is expected, so the use of isopropanol is optional, but it does seem to help the yield.

(7) Place a QIAquick spin column in a 2 ml collection tube.

(8) Apply the solubilized gel sample to the QIAquick column and centrifuge for 1 min. The DNA will bind to the membrane while agarose and small molecules including ethidium bromide will elute into the collection tube. Further washing is unnecessary for the reactions undertaken in the next few experiments.

(9) Decant and discard the flow-through.

(10) Place the QIAquick column back in the same collection tube. Add 0.75 ml (750 µl) QIAgen kit buffer PE to the column and centrifuge for 1 min.

(11) Decant and discard the flow-through. Replace the collection tube and centrifuge for an additional 1 min. The eluent must be removed from the collection tube to prevent any ethanol from vaporizing back into the membrane. The additional centrifugation will remove residual ethanol that was present in buffer PE. Ethanol would inhibit the ligation reaction in Experiment 7.

(12) Place the spin column into a clean, sterile 1.5 ml microcentrifuge tube.

(13) Add 50 µl QIAgen kit buffer EB (10 mM Tris–HCl, pH 8.5) or 50 µl H_2O to the center of the membrane and centrifuge the column for 1 min (for increased DNA concentration add 30 µl EB to the center of the membrane and let it stand for 1 min before centrifuging). No EDTA should be used. It will inhibit subsequent enzymatic reactions.

Preparation for experiment

(1) See Experiment 4.

(2)   6× AGE loading buffer. See Experiment 4.

References

Barzilai, R. (1973). SV40 DNA: quantitative conversion of closed circular to open circular form by ethidium bromide-restricted endonuclease. *J. Mol. Biol.* **74**, 739–742.

Osterlund, M., Luthman, H., Nilsson, S.V. and Magnusson, G. (1982). Ethidium-bromide-inhibited restriction endonucleases cleave one strand of circular DNA. *Gene* **20**, 121–125.

Sambrook, J. and Russell, D.W. (2000). *Molecular Cloning: a Laboratory Manual*. Cold Spring Harbor ,NY: Cold Spring Harbor Press.

Thorne, H.V. (1967). Electrophoretic characterization and fractionation of polyoma virus DNA. *J. Mol. Biol.* **24**, 203–211.

EXPERIMENT 7

(a) Determination of DNA Concentrations
(b) Ligation of Insert into Vector

Introduction

Determination of DNA concentration

It is important to know how much DNA you have in a sample in order to perform reactions with it in the proper proportions. Nucleic acids can be quantitated in a variety of ways. The most reliable and quantitative is by absorption spectroscopy. In this technique, absorption is measured at several different wavelengths to assess both purity and concentration. Very pure nucleic acids have an absorbance peak at 260 nm and their concentration can be determined for very small quantities. However, while nucleic acids have an absorption peak at 260 nm, proteins have an absorption peak at 280 nm. Nucleic acids contaminated with protein will have an enhanced peak at 260 nm due to overlap of the protein absorption peak. It is therefore advisable to read absorption of the sample at both 260 and 280 nm. The $A_{260/280}$ ratio can be used as an indicator of purity. If proteins are present, the 280 reading will be larger than if they were not present and the 260/280 ratio will be correspondingly reduced. Particulates can be detected by looking at the 325 nm reading. Absorbance of aromatics such as phenol which might contaminate the nucleic acid preparation will be detected at 230 nm. Very pure DNA should have a 260/280 ratio of 1.9 or better, although 1.7 is acceptable for many applications.

There are several types of spectrophotometers available. A small single-beam spectrophotometer for which a single cuvette is used and readings recorded manually is commonly found in student laboratories. A scanning spectrophotometer will scan through from 230 to 325 nm, showing relative absorbance peak heights. A double-beam spectrophotometer is able to measure absorbance at different wavelengths in two cuvettes and compare the readings, which greatly simplifies the operation.

In addition to the quantitative spectrophotometric method, the ethidium bromide dot method can be used for estimation of low concentrations of DNA (see Experiment 3). The method cannot determine purity but is inexpensive, specific for nucleic acids and rapid, although somewhat subjective. It is particularly useful when very small amounts of sample are available.

Ligation

After quantitation of your digested and purified target gene insert and pET28b vector, the insert will be ligated into vector. The resulting recombinant plasmid, now carrying the target gene, will be ready for transfer into a host bacterial cell. The pET28b map is shown in Figure 5.1. Note that the position and orientation of the gene insert is dictated by the use of two different restriction sites. Ligation requires the enzyme DNA ligase, which catalyzes the formation of a phosphodiester bond between the 5′ phosphate and 3′-hydroxyl termini in duplex DNA. For sticky ends, an insert/vector ratio of 6:1 is normally used, but higher ratios are common.

Materials and Methods

Materials

Concentration determination

♦   UV–vis spectrophotometer

♦   Matched quartz semi-micro (1 or 0.1 ml) spectrophotometer cuvettes (1 cm path)

♦   UV transilluminator

♦   Goggles

♦   Gloves

♦   P-20 Pipetman and sterile white or yellow tips

♦ DNA standards (calf thymus or λ DNA): 0, 1, 2.5, 5, 7.5, 10 and 20 ng/μl

♦ Ethidium bromide test solution: 1 ng ethidium bromide/μl nuclease-free water or 10 mM Tris−HCl

♦ Purified double digested LTB gene insert (Idd) and pET28b vector (Vdd) from Experiment 6

♦ Plastic cling film

Ligation

♦ Thermal cycler

♦ 200 μl dome-top PCR tubes

♦ T4 DNA ligase kept on ice

♦ 10× ligase buffer supplied with enzyme (see Preparation for Experiment)

♦ 100× or 50× acetylated BSA

♦ Quantitated digested insert (Idd)

♦ Quantitated digested pET28b (Vdd)

Method

Rapid estimation of DNA concentrations using ethidium bromide

See Ausubel *et al.* (1999), and more detailed description and preparation of reagents in Experiment 3.

(1) For a standard curve, make a row of seven spots, 4 μl each, of a 1 μg/ml ethidium bromide solution on a piece of plastic film placed on a UV transilluminator. Add 4 μl of previously prepared (Experiment 3) DNA standard solutions (one standard solution per spot) and mix gently by pipetting up and down two or three times.

(2) Place two 4 μl spots of 1 μg/ml ethidium bromide solution next to the row of standards. Add 4 μl of sample Idd DNA to one spot and 4 μl sample Vdd DNA to the other. Mix gently.

(3) Turn on the UV transilluminator and estimate the unknown DNA concentrations by comparison to the fluorescence of the standards.

Spectrophotometric concentration determination

(1) Pipet 1 ml Tris buffer (10 mM Tris, pH 8.5) into a semi-micro (0.1 ml) or a 1 ml quartz cuvette. This will be the Tris blank. Wipe the cuvette with a KimWipe and place it into the spectrophotometer. Set spectrophotometer to 260 nm. The procedure given here is for zeroing against an air reference and requires only two matched quartz cuvettes, one for zeroing and one for sample.

(2) Zero the spectrophotometer. If the instrument will not zero against air e.g. an empty cuvette, place a cuvette containing distilled water into the holder and zero it against water. Remove the cuvette, replace the water with Tris buffer, wipe the cuvette with a KimWipe and record the absorbance. The absorbance of the Tris blank must then be subtracted from all subsequent readings.

(3) Remove the buffer blank cuvette and replace it with a matched cuvette containing 5 μl of sample added to 95 μl 10 mM Tris (or 50 μl sample + 950 μl 10 mM Tris if you are using a 1 ml cuvette: *if that is your total sample, do not use this method, use the ethidium bromide method above*). Mix by inversion by covering the cuvette with Parafilm, then placing your forefinger firmly on the Parafilm over the mouth of cuvette, your thumb on the base and inverting two or three times. Wipe with a KimWipe. Place the sample cuvette in the spectrophotometer and record the absorbance.

◆ Calculate the apparent amount of DNA in the diluted sample;

◆ Calculate the dilution factor of your sample;

◆ Calculate the apparent amount of DNA in the original sample.

(4) Keep the sample in its cuvette. Remove the sample cuvette and replace it with the Tris blank cuvette, wiped with a KimWipe. Re-zero at 280 nm as described in steps 1 and 2. Re-measure the sample at 280 nm as described.

(5) Calculate the $A_{260/280}$ ratio. This has traditionally been a measure of the purity of the preparation, since protein absorbs at 280 nm, and the protein peak can add into the 260 nm peak, giving a false idea of the DNA concentration. A ratio of 2.0 can be obtained for highly purified calf thymus DNA. However,

recent work indicates that nucleic acids absorb so strongly at 260 nm that only a significant level of protein contamination will cause a change in the absorbance ratio (see Sambrook and Russell, 2000, Vol. 3, p. A8.21 for a discussion of this topic). Calculating a precise concentration by this method may therefore be problematic.

Calculate Sample Concentration

The calculation is based on the Beer–Lambert law, $A_\lambda = \varepsilon_\lambda Cl$ where A_λ is absorbance at a particular wavelength, ε_λ is extinction coefficient at that wavelength and l is pathlength in cm. C is the concentration of DNA. More recently, the law has been formulated as $A = abc$, where A = absorbance, a = molar absorptivity (a more descriptive term for extinction coefficient), b = path length and c = concentration, assuming reasonable purity.

To determine concentration, you can use the following equation (from Ausubel *et al.*, 1999, pp. A3–11) for dsDNA. The equation assumes a 1 cm path length cuvette and neutral pH. The molar extinction coefficient at 260 nm for dsDNA is 6.6.

C (pmol/μl) = $A_{260}/(13.2 \times S)$ where S = the size of the DNA in kb (400 bp = 0.4 kb) (Eq. 7.1)

If you wish to calculate concentration, the extinction coefficient at 260 nm for dsDNA is 20/(mg/ml)/cm meaning an absorbance of 20 at 260 nm corresponds to 1 mg dsDNA/ml. Thus an absorbance of 1 at 260 nm corresponds to 50 μg/ml. The extinction coefficient in μg/ml is 0.020/(μg/ml)/cm and therefore C (μg/ml) = $A_{260}/0.020$.

Table 7.1

Nucleic acid	Concentration to give $A_{260} = 1.0$
dsDNA	50 μg/ml
ssDNA	33 μg/ml
RNA	40 μg/ml

For approximately 28 μg/ml dsDNA, $A_{260} = 0.56$ and $A_{280} = 0.28$ ($A_{260}/A_{280} = 2.0$) can be obtained for highly purified calf thymus DNA in Tris buffer.

Example

Given that: for a sample absorbance of 0.56 the concentration is 28 μg/ml (ng/μl) calculated from the above data; and an average base pair has a molecular weight of 660;

then for an insert (I) of 400 bp the molecular weight of the insert is 400×660 g/mol $=$ 264 000 g/mol $=$ 264,000 ng/nmol.

To calculate concentration, divide 28 ng/µl by 264 000 ng/nmol. You would have 0.0001 nmol/µl $=$ 0.1 pmol/µl (100 fmol/µl) and you would require 3 µl of insert for the ligation mixture.

If the measured concentration is 100 ng/µl, divide by 264 000 ng/nmol $=$ 0.000379 nmol/µl $=$ approximately 0.4 pmol/µl $=$ 400 fmol/µl, and you would need 1 µl.

Calculations

After determining the absorbance or direct concentration, do the calculations. The insert (Idd) encodes 103 amino acids (3 bp/amino acid) plus 21 amino acids ($\times$3 bp) in the leader sequence and 3 bp for each overhang after digestion with restriction endonuclease. Calculate the amount based on the total number of base pairs in the insert. The pET28b plasmid (Vdd) itself is 5368 bp. We have excised 98 bp. Calculate the amount of digested vector based on the *net* size, in base pairs of the plasmid.

(a) The digested insert (Idd) has _____ bp and the digested vector (Vdd) has _____ bp.

(b) The amount of digested insert (Idd) required for a 10-fold excess would be _____ µl.

(c) The amount of digested vector (Vdd) required would be _____ µl.

Method: ligation of LTB gene insert Idd into pET28b vector Vdd to form a recombinant construct

Choose volumes of vector and insert according to their concentrations. The directions given are for a final volume of 20 µl. Approximately 50 ng (0.15–0.3 pmol $=$ ~200 fmol) of a 400 bp insert fragment is used with 50–100 ng (0.015–0.03 pmol $=$ ~15–30 fmol) of pET vector (approximately a 10:1 molar ratio of insert to vector) in a volume of 20 µl.

Idd + Vdd

(1) Mix in a 200 µl PCR tube:
    2 µl 10× ligation buffer or 4 µl 5× ligation buffer (see below for recipes)

x µl prepared insert Idd (0.2 pmol, e.g. 1 µl of 50 ng/µl insert)

y µl prepared vector Vdd (0.03 pmol, e.g. 2 µl of 50 ng/µl vector)

z µl nuclease-free (autoclaved) water to bring to a final volume of 20 µl

1 µl T4 DNA Ligase (two to three units).

Note: ligase must be kept ice cold at all times.

(2) Incubate overnight at 16°C or, depending on supplier, 20 min to 2 h at 37°C or 1–3 h at 25°C. In all cases, a thermal cycler can be used to maintain temperature. Alternatively, a water bath in a cold room can be used for the 16°C incubation. A water bath in a heat-block or at room temperature can be used for the other temperatures.

(3) Set up a control reaction to check for non-recombinant background and possible single digestion. Water is substituted for insert.

Water (W)+Vdd:

2 µl 10× ligation buffer or 4 µl 5× ligation buffer (see below)

x µl nuclease free distilled water substituted for insert

y µl prepared vector (Vdd) (0.03 pmol)

z µl water to bring to a final volume of 20 µl

1 µl T4 DNA Ligase (two to three units).

(4) Incubate overnight at 16°C or, depending on supplier, 20 min to 2 h at 37°C or 1–3 h at 25°C.

(5) After incubation it is imperative to inactivate the enzymes and other proteins in the mixture by heating the reaction mixture at 65°C for 5 min prior to storing in the freezer. Failure to do this may result in poor or no transformation success.

Buffer final concentrations, depending on the enzyme supplier, are: 20–50 mM Tris–HCl (pH 7.6); 10 mM $MgCl_2$; 1–10 mM ATP; 1–10 mM DTT (dithiothreitol); and 25 µg/ml acetylated BSA.

Preparation for experiment

(1) DNA standards: see Experiment 3.

(2) Ethidium bromide test solution: see Experiment 3.

(3) Ligation buffers: Sigma 10× ligase buffer supplied with the T4 DNA ligase kit contains 250 mM Tris–HCl, pH 7.8, 100 mM $MgCl_2$ and 10 mM DTT, supplemented with ATP and PEG-8000, also supplied. Gibco-BRL 5× ligase reaction buffer, supplied with the T4 DNA ligase kit, contains 250 mM Tris–HCl, pH 7.6, 50 mM

MgCl$_2$, 5 mM ATP, 5 mM DTT and 25% (w/v) PEG-8000. Promega 10× ligase reaction buffer supplied with the T4 DNA ligase kit contains 300 mM Tris–HCl, pH 7.8, 100 mM MgCl$_2$, 10 mM ATP and 100 mM DTT. New England BioLabs 10× ligase reaction buffer supplied with the T4 DNA ligase kit contains 500 mM Tris–HCl, pH 7.5, 100 mM MgCl$_2$, 10 mM ATP, 100 mM DTT, 250 µg/ml BSA.

References

Ausubel, F. M., Brent, R., Kingston, R. E., Moore, D. D., Seidman, J. G., Smith, J. A. and Struhl, K. (1999). *Short Protocols in Molecular Biology*, 4th edn. New York: John Wiley, pp. 2–24.

Sambrook, J. and Russell, D. W. (2000). *Molecular Cloning: a Laboratory Manual*. Cold Spring Harbor, New York: Cold Spring Harbor Press.

EXPERIMENT *8*

Transformation of Host Cells by Construct

Introduction

The discovery that mating and DNA transfer occurs between many kinds of organisms, including bacteria, and that this direct transfer can be manipulated, has profoundly altered the course of biological research. There are a variety of ways DNA can be transferred between bacteria, producing a heritable change in the recipient organism.

Conjugation

An 'F' plasmid, containing genes on a linear DNA strand and residing in the cytoplasm of a 'donor' (male) strain of *Escherichia coli* is transferred through a tubular appendage called a pilus to a 'recipient' (female) strain. The recipient can now become a donor.

Hfr conjugation

'High frequency conjugation': in this case the F plasmid has become integrated into the genome (chromosome) of the donor strain. Transfer to the recipient *Escherichia coli* strain is accomplished from this location and often includes transfer of some genomic genes from the donor strain as well as the genes on the F plasmid.

Transduction

A bacteriophage (virus that infects bacteria) mediates the transfer of DNA by attaching to the bacterium and injecting its DNA into the bacterial cell. The phage DNA has instructions for making more phage and may also include genes obtained from other organisms. Both types of genes may be expressed by the host cell.

Transformation

Here naked DNA is taken up directly by the host cell. The DNA may come from a number of different sources and usually contains only a few genes. The process has been observed in nature and was the first type of manipulation used to show that DNA is genetic material and can produce heritable changes, in 1944 by Oswald Avery, Colin Macleod and Maclyn McCarty at Rockefeller Institute in New York (Avery, Macleod and McCarty, 1944).

Cloning can be defined as the amplification of the gene of interest as a heritable change in identical cells descended from a single transformed cell. The transformation technique used in this experiment will transfer the plasmid construct pET28b with LTB gene insert (Experiment 7) to a non-expression bacterial host cell for amplification of the plasmid and inserted gene. This cloning step is performed prior to expression because it provides large amounts of plasmid construct for sequencing and subsequent transformation into expression hosts. It is also a way to identify problems like vector stability, amplification or bacterial resistance that may arise during later expression steps. The construct is generally stored in a non-expression host for convenience and stability. For the purpose of cloning and storage, the non-expression host *E. coli* strain BL21 will be used, which does not contain the λDE3 plasmid that carries the gene for T7 RNA polymerase. Background expression is minimal in the absence of this polymerase, since the target gene is under control of the T7 promoter. More information on the bacterial genotype can be found in the *pET System Manual*, available from the Novagen website, www.novagen.com.

In order to transform bacterial cells, the bacterial cell must be made *competent* to take up DNA. This is accomplished by weakening its cell membrane. There are two common methods. The simplest is incubation of the cells in $CaCl_2$ followed by mild heat shock when the temperature is sharply increased from 4°C on ice to 41°C, then the cells are quickly chilled. This method is gentle on the bacterial cell but DNA is transferred into the bacterial host less efficiently. Only about 10% of the cells are actually transformed (have taken up the naked foreign DNA). Alternatively, electroporation may be used. An electric current from an electroporation apparatus is passed through the solution containing the bacterial host cells and plasmid vector DNA. The current causes transient, rapid opening of channels and pores in the plasma membrane of the bacterial host cells, and plasmid vector DNA is pulled in through these transient openings. The method is very efficient at getting DNA into the bacterial host cell, but kills about 40–80% of the cells exposed to the current. Of the remaining 20–60%, most have taken up naked DNA.

Overall, the number of transformed cells is generally larger using the electroporation technique than using $CaCl_2$ and heat shock. The $CaCl_2$/heat shock method gives reasonable results and is much less expensive to use. Competent cells can be purchased, or can be made competent just prior to use.

Materials and Methods

Materials

Transformation

♦ Thermal cycler, water bath or heat block with water in block wells set at 42°C

♦ Shaker water bath set at 37°C

♦ Ice/ice bucket and individual containers

♦ P-20 and P-200 Pipetmen with sterile white or yellow tips

♦ Competent *E. coli* BL21 cloning host cells (20 μl aliquots in sterile 0.5 or 1.5 ml microcentrifuge tubes)

♦ Ligation mixture (Idd+Vdd) containing LTB insert-pET28b construct from Experiment 7

♦ Ligation control mixture (W+Vdd) containing pET28b doubly digested (optional)

♦ Test plasmid control, supplied with competent cells (Novagen, Madison, WI)

♦ SOC media (100 μl/20 μl cell aliquot), supplied with competent cells (Novagen, Madison, WI) see the recipe in Preparation for Experiment

Plating

♦ 37°C incubator

♦ LB-Kan30 agar plates

♦ Glass hockey stick (spreader) or pre-sterilized Lazy-L spreaders (Sigma catalog no. Z37 677-9)

♦ 95% ethanol and Bunsen burner for glass spreaders

Method

Transformation *

(1) Aliquots of 20 µl competent BL21 cells in a sterile tube have been prepared. *Keep them cold, on ice*.

(2) (a) Add 1 µl of the previously purified ligation mixture (pET28LTB) to the ice-cold BL21 cells and gently stir with the pipet tip to mix. Do not be rough as the cells are very fragile!

   The following controls should be done for the class by one or two teams each:

   (b) add 1 µl of the control mixture (W+Vdd, double digested pET28b ligated with water substituted for insert) to a tube containing 20 µl iced BL21 cells;

   (c) for a further control, to judge transformation efficiency, add 1 µl of test plasmid to a tube containing 20 µl iced BL21 cells

(3) Store the cell mixtures on ice for 5 min to allow the cells to recover from mixing. Preheat the thermal cycler or heat-block to 42°C.

(4) Wipe the bottom of the tubes. Holding the tube by the top, dip the lower half of the tube in the 42°C water bath for *exactly* 30 s to heat-shock the cells.

(5) Remove the tubes and immediately place them in ice up to their lids. Leave them for 2 min to recover.

(6) Add 80 µl of room-temperature SOC media to each tube. Replace the tubes in ice until ready to incubate. SOC medium contains tryptone 20 g/l; yeast extract 5 g/l; NaCl 0.5 g/l; and 20 mM glucose. It allows the cells to recover from the shock of transformation in a comfortable growth environment.

(7) The mixture should be incubated at 37°C with shaking (250 rpm) for 1h prior to plating to allow the cells to recover from shock. This is called an 'outgrowth incubation'. It can be conveniently done by placing the tubes in empty 15 ml plastic centrifuge tubes or 13×100 mm glass test tubes in a rack in the shaker. The snap-caps will keep the transformation mixture vertical and prevent the tubes from falling in.

(8) Place the agar plates, cover side down in the 37°C incubator to warm and dry.

---

*Reproduced by permission of Novagen.

Plating on selective media

The construct, if successfully ligated, is a newly formed recombinant plasmid that will be called pET28LTB hereafter. It carries the gene for antibiotic resistance (kan). The protein expressed by this gene is an enzyme that degrades the antibiotic Kanamycin. Thus, only cells transformed by the construct are selected (can survive) on the antibiotic-containing media. The plates are usually incubated for 24 h at 37°C then removed from the incubator, taped shut, and stored in the refrigerator to limit occurrence of spontaneous mutants. Although cells transformed by pET28b plasmid without insert grow well in 24 h, the presence of the pET28LTB construct seems to slow the growth of transformants. Therefore it may be necessary to give the cells as much as 48 h incubation for colonies to grow to a satisfactory size. This should not result in satellites on Kan30 plates, as might have occurred if Ampicillin or Carbenicillin had been the selective antibiotic (see notes on selective media, Experiment 2).

(1) Pick up the lid of the plate with one hand and pipet 50 μl of the SOC–cell mixture onto the agar surface. Replace the lid.

(2) If using the Lazy-L spreader, remove it from its sleeve and use it to lightly spread the cells/media mixture over the surface of the agar.

   Otherwise:

(3) Turn on the burner and make a blue flame. *Keep the beaker or Petri dish containing 95% ethanol away from the flame.*

(4) Immediately remove the bent glass rod ('hockey stick') spreader from the ethanol bath and sterilize it by holding it in the flame. The alcohol will burn off.

(5) *Do not return the hot rod to the ethanol bath.* When the flames have died out, open the lid of the plate containing the cells and rest the spreader on the edge of the agar to cool it a bit on the sterile agar surface. A hot rod will kill the cells.

(6) Align the spreader on the agar, perpendicular to the edge the plate, and lightly spread the cells evenly over the surface by rotating the plate. Do not press down with the spreader. Quickly cover the plate.

Samples

♦ Plate 50 μl pET28LTB-BL21 mixture on one plate of selective agar containing Kanamycin.

♦ Plate 25 µl pET28LTB–BL21 mixture on another selective agar plate.

♦ Plate 50 µl control W+Vdd–BL21 mixture.

♦ For the test plasmid (T), pipet 20 µl fresh of SOC onto the center of a fresh LB-Kan30 plate. Add 5 µl test plasmid to the center of the SOC puddle. Using the same pipet tip, pick up 5 µl SOC from the edge of the puddle and dispense it into the center of the puddle. This will rinse the tip and mix the cells. Spread with a sterile spreader as previously.

(7) Allow the plates to sit on the bench top for 5–10 min, lids closed, to allow the plating media to be absorbed into the agar.

(8) After the plates have been upright for several minutes, turn them over and label them on the bottom around the edge noting: (a) your name; (b) the name of the construct; (c) the host cell strain; and (d) the date of plating.

(9) Place the plates, inverted, in the 37°C incubator. The plates are kept inverted so that the condensate does not drip onto the surface and spread the colonies over the agar.

(10) The next day: come in and check the number of colonies (transformants) on your plates. Seal the plates with a strip of Parafilm and store, inverted, in the refrigerator (4°C). If no colonies appear, let the plates incubate for 24 h more.

Note

The number of transformants is variable. If there are so many individual colonies that you count over 50 individual colonies note 'over 50'. If there are so many that individual colonies run into one another and form a lawn, check the 25 µl plate. If the 25 µl plate is overrun, and an individual colony cannot be picked, replate by picking up a tiny bit of lawn with a sterile loop and streaking a fresh LB-Kan30 plate then incubating it at 37°C overnight.

If there are no colonies or only one or two colonies per plate, these may be 'background transformation' meaning that the plasmid did not re-ligate (with or without insert), but some few uncut or singly cut, re-ligated plasmids have transformed the host cells. There are several possible reasons for this last result:

(1) The ligase may be inactive. Ligase is notoriously sensitive to freezing and thawing; try the reaction again with a fresh batch of enzyme.

(2) The vector may have been cut by both restriction endonucleases but the insert does not have compatible sites or is not present. The vector cannot re-ligate unless it

contains insert because the double-digested vector has incompatible ends. To test, add ligase to Vdd and check it by AGE. If the vector has incompatible ends, it should still be linear. To check the insert, do a ligation reaction without vector present and check the mixture by AGE, compared to un-ligated insert. If your ligation mixture produces multiple bands because individual linear products have been ligated together, it suggests that the ends are properly cleaved. If you see only dimers in the ligation mixture the result suggests that only one end of the insert has been cleaved. Recut to determine which end has not been cut.

(3) No vector was present. The insert alone does not contain the antibiotic resistance gene.

The test plasmid plate should have many individual colonies (~100). If not, there is a problem with your transformation technique or reagents. If there are no colonies on either plate, but other students do, cells can be obtained from another student for the next experiment.

In general the pET28LTB plate should have more colonies than the control plate, but an equal number does not mean that the insertion was unsuccessful. On the other hand, if the vector was cut by only one of the two restriction endonucleases, the insert will not be present but the vector will have compatible ends and can re-ligate, transform and confer resistance on the host cell. It is therefore necessary to grow several individual colonies and check (by AGE) for presence of insert before proceeding.

Preparation for experiment

(1) Aliquot purchased *E. coli* BL21 cloning host cells (Novagen Inc., Madison, WI, USA, catalog no. 69449-3). Cells should be stored at $-70°C$. Handle only the very top of the tube and tube cap to prevent warming. To mix the cells, flick the tube, *never vortex*. Cells can be thawed once to aliquot them then re-frozen at $-70°C$. Additional freezing and thawing will seriously compromise their ability to be transformed. To aliquot from a 0.2 ml stock, place 10 sterile 0.5 or 1.5 ml tubes in ice to chill. Remove the 0.2 ml stock from the $-70°C$ freezer and immediately place on ice. To dispense, remove the tube of cells from ice and flick once or twice to mix prior to opening the tube. Remove a 20 µl aliquot from the middle of the cells, and replace the stock tube on ice immediately. Place the aliquot into the bottom of the pre-chilled tube, mix by pipetting up and down *once*, then quickly close the tube and replace in ice. After all the aliquots are taken, return any unused tubes to the $-70°C$ freezer before proceeding with the transformation.

(2) To prepare competent *E. coli* cells. Rapid one-step preparation (Ausubel *et al.*, 1999, pp. 1–28). A longer protocol is also available. Prepare an overnight culture of *E. coli* BL21 cells in LB broth (Experiment 5). Dilute it 1:100 in fresh LB broth. Grow at 37°C to an OD_{600} of 0.3–0.4. Add 1 vol. of ice-cold 2× TSS (see recipe below) and gently mix on ice. Small aliquots may be frozen at −70°C for future use. They should be thawed on ice immediately before use.

2× TSS (transformation and storage medium)

Dilute sterile (autoclaved) 40% polyethylene glycol (PEG) 3350 to 20% PEG in sterile LB medium containing 100 mM $MgCl_2$. Add dimethylsulfoxide (DMSO) to 10% (v/v) and adjust the pH to 6.5.

(3) SOC medium ('save our cells'?): 20 g tryptone; 5 g yeast extract; 0.5 g NaCl; 20 mM glucose; distilled water to 1l. Autoclave to sterilize.

An alternative recipe for SOC (Ausubel *et al.*, 1999) is: 0.5% (w/v) yeast extract; 2% (w/v) tryptone; 10 mM NaCl; 2.5 mM KCl; 10 mM $MgCl_2$; and 20 mM glucose.

References

Ausubel, F. M., Brent, R., Kingston, R. E., Moore, D. D., Seidman, J. G., Smith, J. A. and Struhl, K. (1999). *Short Protocols in Molecular Biology.* New York: John Wiley.

Avery, O. T., Macleod, C. M. and McCarty, M. (1944). Studies in the chemical nature of the substance inducing transformation of pneumococcal types. Induction of transformation by a desoxyribonucleic acid fraction isolated from *Pneumococcus* type II. *J. Exp. Med.* 79, 137–158.

EXPERIMENT 9

(a) Colony-pick PCR to Check for Presence of Insert
(b) Streak Plates with Potential Constructs
(c) Inoculate Cultures for Plasmid Isolation

Introduction

After cells are transformed and grown on selective media it is still necessary to verify that the colonies contain the construct (LTB insert-pET28b vector) rather than just the pET28b plasmid vector, since the presence of the plasmid, whether or not it contains insert, confers antibiotic resistance on the cells. There are several methods that can be used to test for the presence of insert in successfully transformed cells. One technique, used for large (700 bp) inserts, uses a small amount of culture grown from a colony. The culture is subjected to agarose gel electrophoresis without plasmid isolation. Although genomic DNA and RNA appear on the gel, the plasmid can also be detected and its size compared to plasmid without insert, assuming the plasmid is present in sufficiently high copy numbers. Other methods require an inoculum to be grown up in selective media from a colony then the plasmid isolated from the culture. The purified plasmid can then be electrophoresed to compare the size of the isolated plasmid with plasmid that does not contain insert. One may also do a restriction digest to free the insert from the vector, then electrophorese to visualize the vector band and the insert band, if present.

In addition to these 'after the fact' methods, a clever use of the β-galactosidase gene in a complementation system can be used. The system is called α-complementation. A special vector carries a short segment of *E. coli* DNA encoding the regulatory sequences and coding information for the first 146 amino acids of the β-galactosidase gene, in which a polycloning site has been embedded. The complementary bacterial host cells express the carboxy-terminal portion of the enzyme. Neither of the fragments is enzymatically active, but when the plasmid vector is introduced into the host bacterial cell, production of the N-terminal portion of the β-galactosidase gene present on the vector plasmid is induced by the lactose analog isopropylthio-β-D-galactoside (IPTG). The two portions of the expressed β-galactosidase protein can associate and form an active enzyme. Use of the substrate X-Gal (5-bromo-4-chloro-3-indolyl-β-D-galacto-side), an analog of the lactose substrate, permits visualization of enzymatic activity. When the X-Gal is cleaved by active β-galactosidase the chromophore portion of X-Gal is released as a deep blue precipitate. If the vector, without insert, transforms a host cell, the cells, grown on selective agar, will produce a deep blue color when the two parts of β-galactosidase function together. On the other hand, if the target gene is inserted into the cloning site within the β-galactosidase gene in the vector, the N-terminal portion of the enzyme is no longer able to function and white colonies result. Thus, one can know immediately that transformation of a recombinant construct has occurred (Sambrook and Russell, 2000, p. 1.150). The method, however, is not infallible, and it does not validate the size of the insert. It is best, even when using X-Gal, to verify insert size.

The method chosen for this experiment allows use of a very small amount of material from a single colony to amplify by PCR a section of the plasmid that should contain LTB insert. The amplified fragment is subjected to agarose gel electrophoresis and compared to a control PCR product from intact pET28b that does not contain insert. Colony-pick PCR is useful because it does not depend on plasmid copy number and it gives a very good result for smaller inserts (200–400 bp). More importantly, it results in a PCR product that has a measurable size difference so the presence of an insert of known size can be observed. The LTB gene has been inserted between the *Bam*HI and the *Nco*I sites on the pET28b plasmid vector. The fragment chosen for amplification lies between the T7 promoter sequence and the T7 terminator sequence on pET 28b and contains the restriction sites used for insertion of the target gene (see vector map, Figure 5.1). The forward primer (T7 promoter) is 5′TAATACGACTCACTATAGGG3′ and the reverse primer (T7 terminator) is 5′GCTAGTTATTGCTCAGCGG3′. If no target gene has been inserted the fragment will be 360 bp. If the target gene has been inserted the fragment will be 633 bp total size, accounting for a 98 bp excision and a 375 bp LTB insert.

One or more transformed colonies is chosen. Each colony will be picked with a sterile toothpick, or a sterile pipet tip. The colony pick is first swished into sterile water to provide template DNA for the PCR. Next, the pick is used to streak a selective agar plate for storage of the putative construct. Finally, the same pick will be used to inoculate a small liquid culture to produce bacteria containing sufficient template plasmid for sequencing. The series of manipulations makes very efficient use of a small single colony of transformed bacteria.

Materials and Methods

Materials

♦ Thermal cycler

♦ P-20 Pipetmen and sterile white or yellow tips

♦ Sterile 200 μl PCR tubes

♦ Fresh LB-Kan30 plate divided into eight sectors

♦ 15 ml sterile centrifuge tubes containing 3 ml LB-Kan30 broth (one per colony picked)

♦ 10× Mg-free *Taq* buffer, supplied with enzyme and kept on ice

♦ *Taq* DNA polymerase, diluted 1:10 (to 5 U/ml) in 1× *Taq* buffer and kept on ice at all times

♦ 25 mM MgCl$_2$

♦ 2.5 mM dNTP mix [alternatively a PCR master mix (2×; Promega, Madison, WI) may be used, containing *Taq* DNA Polymerase (50 U/μl in reaction buffer pH 8.5); 400 μM each of dATP, dGTP, dCTP and dTTP; and 3 mM MgCl$_2$ kept on ice at all times]

♦ Plates of transformants from Experiment 8

♦ PET 28b plasmid DNA or a colony containing the plasmid

♦ Nuclease-free water (autoclaved distilled water)

♦ Forward primer (T7 promoter primer) 10 pmol/μl (10 μM)

♦ Reverse primer (T7 terminator primer) 10 pmol/μl (10 μM)

♦ 6× gel loading buffer (AGE-LB)

Method: pick; streak; inoculate

(1) Pipet 20 μl nuclease-free water into a sterile PCR tube (one tube per colony to be tested). Number the tubes.

(2)   On the outside bottom of the plate of transformants, circle a colony and label it with a number. Open the lid with one hand and use the other to pick the colony from the plate using a sterile P-200 tip or sterile toothpick. *Wear gloves.* Swish the tip back and forth two or three times in the numbered tube containing 20 μl nuclease-free water. The tip, attached to a Pipetman, can be used to do the pick, but the Pipetman barrel must be cleaned by rinsing it with 95% ethanol before attaching the tip. If there are early colonies and ones that show up in the second 24 h of incubation, choose one colony from each subset.

(3)   Next, use the same tip to lightly streak a quadrant on a sectored fresh Kan-30 plate. When all colony replicates have been made, the plate will be incubated overnight at 37°C, then stored in the refrigerator (4°C).

(4)   Now drop the same tip into a 15 ml centrifuge tube containing 3 ml LB-Kan30 broth. The cap of the tube should be removed, the top of the tube introduced to the flame very briefly and the tip dropped into the broth. Flame again, quickly (be careful – plastic melts!) and replace the cap, screwing it on *loosely* but securely (tighten then turn back a quarter or half a turn). The caps must be just loose enough to allow some air to enter so the bacteria can grow efficiently, but not so loose that they will fall off. The top of the tip should be above the broth, otherwise there may be contamination from the gloves or Pipetman barrel. Label the tube with the experiment number, the colony number, name/ID, and the date. Inoculated tubes will be incubated overnight in a 37°C shaker bath. After overnight incubation the caps will be tightened and the tubes stored in the refrigerator (4°C). After confirmation of the presence of the LTB insert and its sequence, the tubes will be harvested.

(5)   Repeat this procedure for each colony chosen. At least two construct-transformed colonies should be tested per plate.

Set up and run PCR

(6)   Boil the 20 μl samples in the thermal cycler or a water bath at 100°C for 5 min. The distilled water will help to lyse the cells and boiling denatures the bacterial proteins. Remove the tube and centrifuge for 1.5 min to collect condensate and sediment denatured proteins and membranes. The DNA should remain in solution.

(7)   Set up a PCR reaction in fresh PCR tubes labeled with the colony ID.
Add (total of 20 μl)
2–3 μl boiled DNA template from the middle of the supernatant step (6)
2 μl 10× Mg-free buffer
1.6 μl 2.5 mM dNTP mix

1.2 µl 25 mM MgCl$_2$

1 µl Forward primer (T7 promoter) 10 pmol/µl

1 µl Reverse primer (T7 terminator) 10 pmol/µl

9.2 µl sterile water (or enough to make up to a final volume of 20 µl)

2 µl fresh *Taq* DNA polymerase (5 U/µl)

If kept to a minimum, cell cytosolic components should not interfere with the polymerase.

Alternatively, if using commercially prepared (or previously prepared) 2× master mix, add to a sterile PCR tube (total of 25 µl); 2–3 µl boiled DNA template taken from the middle of the supernatant; 12.5 µl 2× master mix; 6.5 µl nuclease-free water; 1.5 µl F (T7 promoter) primer, 10 µM; and 1.5 µl R (T7 terminator) primer, 10 µM.

(8) Set up a control reaction with 10–50 ng PET 28b plasmid DNA

(9) Run Method 444*Taq*:

     (1) 94°C, 3 min – main denaturation

     (2) 94°C, 30 s – cycling denaturation

     (3) 40°C, 30 s – cycling annealing

     (4) 72°C, 1 min – cycling extension

     (5) 39× to (2) – 40 cycles (steps 2–4)

     (6) 72°C, 5 min – finishing step for full-length extension

     (7) 15°C, hold – stop reaction

     (8) End.

The annealing temperature is 40°C for because the melting temperature of the T7 promoter primer is 46°C. The melting temperature of the reverse primer, T7 terminator sequence, is 58–60°C, so the lower temperature should ensure annealing of both primers. The promoter and terminator sequences are used as priming sites so only that portion of the vector containing (or not containing) the insert can be 'PCRed-out'. The program is set for 40 cycles because there is less concern with errors and more concern for having sufficient material to view on the gel. The 15°C hold is set because is it easier for a thermal cycler to maintain that temperature than 4°C. This reaction takes just over 2 h to run. If necessary it can be run overnight removed and stored next morning.

(10) When the PCR reaction is over add 5 µl 6× AGE-LB to the mixture, mix gently and freeze the samples.

Preparation for experiment

(1) dNTP mix: this is commercially available from several suppliers as a 10 mM solution containing 10 mM dATP, 10 mM dGTP, 10 mM dCTP and 10 mM dTTP. The mix can

be diluted 1:4 with nuclease-free water for use. Also 25 mM mixes are available, and should be diluted to 2.5 mM for use. Separate vials are also available, as 100 mM sets which can diluted to the requisite concentrations.

(2) The forward primer (T7 promoter) is 5′TAATACGACTCACTATAGGG3′ and the Reverse primer (T7 terminator) is 5′GCTAGTTATTGCTCAGCGG3′. These can be obtained from several suppliers including Integrated DNA Technologies (www.idtdna.com).

(3) LB-Kan30 plates; see Experiment 5.

(4) 6× AGE-LB; see Experiment 4.

Reference

Sambrook, J. and Russell, D.W. (2000). *Molecular Cloning: a Laboratory Manual*. Cold Spring Harbor, NY: Cold Spring Harbor Laboratory Press.

EXPERIMENT *10*

(a) Agarose Gel Electrophoresis to Verify Presence of Insert
(b) Purification of pET28LTB using a Plasmid Miniprep Kit

Introduction

Sizing the PCR fragments

A quick gel will be run to determine whether the insert is present. A fragment beginning with the T7 polymerase promoter region and ending with the T7 terminator sequence (Fig 5.1) has been amplified by PCR in Experiment 9. Review the pET28b plasmid sequence to identify the positions of the promoter and terminator sequences and note the insertion position for the LTB gene. You will find that the LTB gene insert should lie between the two T7 sequences. Thus, if the gene is not present, a short 350 bp fragment will have been amplified. If the LTB gene has been inserted, the fragment. will be approximately 630 bp (350 bp less 98 bp excised between restriction sites plus 375 bp LTB gene insert). The difference in size and mobility is obvious on a gel. The W+V fragment and a pET28b template are the controls (no insert) for comparison.

Plasmid purification: plasmid miniprep

After analyzing the gel, cultures of bacteria grown from a colony shown by PCR and AGE to contain the target gene construct will be chosen. The plasmid constructs will be purified from the cell culture for use as template DNA to be sequenced. DNA sequencing will ensure that the gene is present, accurately copied, and in the correct orientation required for proper expression. Further, any stored successfully sequenced DNA can be saved for subsequent transformation of an expression host. A small amount of culture will be stored as a glycerol stock for use as an inoculum to grow more plasmid if it is necessary to obtain additional plasmid for transformation into an expression host.

The alkaline lysis procedure is the most commonly used as a 'miniprep' for isolating and purifying plasmids from bacterial cells. Experiment 2 describes the steps in detail. A miniprep is usually 3–10 ml of cultured cells. Larger amounts of material use the same protocol, but are called 'midiprep' or 'maxiprep', depending on the volume to be processed. A miniprep should yield about 2–2.5 μg total DNA. Briefly, to review, approximately 5 ml of LB broth medium is inoculated with a single bacterial colony previously grown on an agar plate. When the bacteria have multiplied to about 10^6–10^8 cells per ml ($OD_{600} \approx 0.6$–0.8), they are harvested by centrifugation. The bacterial pellet is resuspended, usually in a glucose/Tris/EDTA solution, then NaOH and SDS are added to lyse the cells (alkaline lysis). Potassium acetate (KOAc) is added to neutralize the mixture, and coincidentally, cause precipitation of protein and chromosomal DNA by mixed ion $(Na^+–K^+)SDS$. The supernatant, containing plasmid DNA and some contaminants, is removed, and the plasmid DNA in the supernatant is precipitated by addition of 95% ethanol at room temperature. The solution is centrifuged to pellet the plasmid DNA. The supernatant, containing a variety of contaminating materials, is discarded. The DNA pellet is washed by addition of a volume of 70% ethanol. The tube is centrifuged again and the wash ethanol discarded without disturbing the DNA pellet. The DNA pellet is air-dried or dried under vacuum, then resuspended in a small volume of Tris/EDTA pH 7.4 (TE), Tris pH 8.5 or nuclease-free distilled water (Ausubel *et al.*, 1999; Birnboim, 1983; Sambrook and Russell, 2000, Vol. 1).

Materials and Methods

Materials

AGE

♦ UV transilluminator

♦ Goggles

♦ Latex gloves

♦ Number 30 sunscreen, face mask/face visor

♦ P-20 and P-200 Pipetmen, sterile white or yellow tips

- Horizontal AGE apparatus

- Gel casting platform, gel combs (slot formers), power supply

- 125 ml Erlenmeyer (conical) flask

- Agarose

- 50× TAE or TBE electrophoresis buffer diluted to 1×

- Ethidium bromide stock, 10 mg/ml water

- 6× loading buffer (6× AGE-LB)

- PCR fragments from Experiment 9

- PCR control fragment (W+V from Experiment 9 and/or pET28b PCR fragments)

- 1 kb ladder, PCR markers

- Plastic cling film

Miniprep

- Microcentrifuge

- P-20 and P-200 Pipetmen, sterile white or yellow tips

- Heat block set at 65°C

- QIAprep Spin Miniprep Kit (QIAgen catalog no. 27104)

- 1.5 ml sterile microcentrifuge tubes

- Fine-drawn Pasteur pipet or gel-loading tips for Pipetman

- 3 ml cell cultures from Experiment 9

- 80% (v/v) glycerol, autoclaved

- Nuclease-free water (autoclaved distilled water)

- 70% (v/v) ethanol

Method

AGE

(1) Make 1× TAE buffer (final concentration 40 mM Tris acetate/1 mM EDTA/pH 8.0) by diluting 20 ml of the 50× stock TAE to 1000 ml. Each gel will require 300 ml 1× buffer.

(2) Prepare a 1.3% gel (0.4 g agarose/30 ml 1× buffer). In a 125 ml Erlenmeyer (conical) flask, mix 30 ml 1× TAE buffer with 0.4 g agarose. Boil the mixture in a microwave oven for 1–2 min until it clears or boil it on a hot plate until the solution is clear. Remove and swirl the mixture to mix and de-gas. Let it cool to 60°C for safe handling. When cool enough to handle, add 3 μl ethidium bromide stock (for a final concentration of 0.66 μg/ml). Swirl to mix.

(3) Set up the gel casting platform. Pour the gel slowly and evenly into the center of the platform. Place the comb (well-former) in its slot about 2 cm from the end, using the eight-well side.

(4) After the gel has hardened and become opaque (15–30 min) remove the seals or dams from the platform and remove the comb. Place the platform into an electro-phoresis tank so the wells are closest to the negative electrode. Pour in 250 ml electrophoresis buffer, to cover the gel about 1 mm above its surface.

(5) Load:
well 1, PCR markers (10–15 μl in AGE-LB)
well 2, pET28b control fragment (20–22 μl) in AGE-LB
wells 3–8, colony-pick PCR samples (20–22 μl/well) in AGE-LB.

(6) Electrophorese with 85–90 V for 30–40 min until dye bands separate and yellow band is a half to two-thirds of the way down the gel. These are small fragments and only their sizes need to be compared.

(7) Remove gel to plastic wrap and view on the transilluminator. The distinction between the presence and absence of insert should be clear. Note the colony ID of samples containing insert. Put a check mark or star on the storage plate quadrant for all positive colonies, and on the positive broth cultures.

Plasmid miniprep (QIAprep® Plasmid Miniprep)*

Prepare lysate and make glycerol stock   (1) Place 1.5 ml of a positive culture, judged by the presence of the insert, in a sterile 1.5 ml microcentrifuge tube. Orient the tube centrifuge so the hinge faces inward. Centrifuge for 10 min to pellet the cells.

(2) Discard the supernatant and re-fill the tube with all except 0.3–0.5 ml of the remaining culture. Replace the tube in the centrifuge, hinge in, so the second pellet will sediment on top of the first. Centrifuge for 10 min.

---

*Reproduced by permission of QIAgen.

(3) Add 75 µl 80% glycerol to the 0.3–0.5 ml culture remaining in the 15 ml centrifuge tube. Cap the tube tightly and store it at 4°C; the bacteria do not grow well in the absence of oxygen, nor do they grow at 4°C. Glycerol stock can be frozen (preferably at −70°C) and stored for several months to be used as an inoculum later, should it be needed.

(4) Resuspend the pelleted bacterial cells in 250 µl of QIAgen kit buffer P1 (glucose/ Tris/EDTA containing RNase). Mix well by pipetting up and down.

(5) Add 250 µl QIAgen kit buffer P2 (NaOH/SDS) and gently invert the closed tube four to six times to mix. Be gentle to avoid shearing the chromosomal (genomic) DNA. Sheared chromosomal DNA would be isolated with the smaller plasmid DNA, contaminating your preparation. Do not allow lysis to proceed for more than 5 min.

(6) Add 350 µl of QIAgen kit buffer N3 (KOAc, high salt) to denature and precipitate proteins, cellular debris and SDS. Immediately mix gently by inversion four to six times. The solution will become cloudy.

(7) Centrifuge for 10 min at 7000–10 000 g.

(8) Carefully remove the supernatant with a Pipetman and clean gel-loading tip. Place the supernatant directly into a QIAprep spin column seated in its 2 ml collection tube. The supernatant may also be decanted directly into the spin column.

(9) Centrifuge for 60 s and discard the flow-through.

(10) Wash the column by adding 0.75 ml (750 µl) QIAgen kit buffer PE and centrifuging for 60 s.

(11) Discard the flow-through, then return the column to a 2 ml collection tube and centrifuge for an additional 60 s to remove the last traces of ethanol from the filter.

(12) Place the spin column in a sterile 1.5 ml microcentrifuge tube and label the tube with initials and 'pET28LTB DNA'.

(13) Add 50 µl QIAgen kit buffer EB (10 mM Tris–HCl, pH 8.5) or autoclaved water to the center of the filter membrane. TE cannot be used in fluorescent DNA sequencing protocols such as the ABI Prism PE Biosystems (now Applied Biosystems). DNA to be used for enzymatic reactions and sequencing must be resuspended in water or Tris–HCl, since EDTA may inhibit enzymes that require Mg^{2+} as a cofactor.

(14) Let stand for 1 min, then centrifuge for 2 min.

(15) Label the tube. Freeze at −20°C. DNA in water or Tris buffer is stable stored at −20°C or below.

Preparation for experiment

(1) 50× TAE electrophoresis buffer, pH 8.5 (500 ml); see Experiment 4. 1× working solution is 40 mM Tris-acetate/2 mM EDTA.

(2) 6× AGE gel loading buffer; see Experiment 4

(3) It is useful to aliquot individual portions with a few (2–10) µl extra for each solution used from the kit to avoid contamination of the bulk solution and simplify distribution.

References

Ausubel, F. M., Brent, R., Kingston, R. E., Moore, D. D., Seidman, J. G., Smith, J. A. and Struhl, K. (1999). *Short Protocols in Molecular Biology*. New York: John Wiley.

Birnboim, H. C. (1983). Rapid alkaline extracton method for the isolation of plasmid DNA. *Meth. Enzymol.* 100, 243–249.

Sambrook, J. and Russell, D.W. (2000). *Molecular Cloning: a Laboratory Manual*. Cold Spring Harbor, NY: Cold Spring Harbor Laboratory Press.

EXPERIMENT **11**

(a) DNA Concentration Determination
(b) Ethanol Precipitation of Plasmid DNA
(c) Sequencing Reaction

Introduction

Ethanol precipitation

The target gene is said to have been cloned when a colony of bacteria containing the plasmid construct has been identified and grown up. The next step is to verify that the sequence of the LTB insert is correct and in the correct orientation for transcription. This is accomplished by purifying the plasmid construct DNA and sequencing it or a section of the plasmid that contains the insert. The pET28LTB plasmid DNA will serve as template for a small PCR product (300–700 bp) lying between the T7 promoter and the T7 terminator sequences. This portion of the plasmid contains the inserted LTB sequence. It is necessary to remove every trace of contamination from the template DNA to ensure that the automated fluorescent sequencing will proceed cleanly and provide reliable, creditable data. This is accomplished by precipitating the DNA from the purification step with ethanol, washing the pellet with additional ethanol to remove all traces of contamination, then drying the pellet and resuspending it in a known volume of water.

Concentration determination

Recovery is estimated by using a small sample of the original 50 µl plasmid DNA preparation from Experiment 10 prior to precipitation, then use that information to determine the final volume required for resuspending the plasmid template DNA. Cycle sequencing reactions are made up to a 20 µl volume, and allow for 8–11 µl of DNA template and 1–4 µl primer. The reaction will require 200–500 ng template DNA per reaction mixture, meaning at least 25–30 ng/µl of clean plasmid template DNA will be needed for a successful sequencing reaction. The ethanol precipitation and subsequent resuspension will allow adjustment of the concentration of DNA based on an estimate of the total amount of DNA in the sample. To obtain a more exact estimate 4–5 µl DNA can be run on a thin agarose gel and the amount estimated in ng using mass markers (for example, Mass Ruler, BioRad, catalog no. 170-8207 or 170-8356) in one lane.

Sequencing the construct

DNA sequencing has become relatively routine as rapid, inexpensive techniques have been developed. Recent advances in sequencing have produced nucleotide sequences for entire genomes of many bacteria, fungi, plants and now the human genome. An extensive and interesting discussion of the history of sequencing can be found in Sambrook and Russell (2000, Vol. 2, p. 12.94). Most modern sequencing methods are based on the Sanger dideoxy method in which DNA polymerase makes copies of a template DNA using primers complementary to the beginning of the sequence of interest. Primer extension follows. If the reaction contains an excess of the usual $2'$ dNTPs plus a small amount of $2',3'$ dideoxy nucleotide triphosphates (ddNTPs), then a ddNTP will occasionally be incorporated into the growing chain. When that happens, the chain terminates because no $3'$-OH group is available to act as a nucleophile to form the phosphodiester bond with the incoming dNTP. When the concentrations of template DNA, dNTP and ddNTP are optimized, the products of the reaction mixture are a series of extended primers that have been lengthened by $n+1$ nucleotides.

A variation of the Sanger method called ABI Prism™ BigDye™ Terminator Cycle Sequencing will be discussed in this experiment. Instrumentation and reagents come from Applied Biosystems. The sequencing reaction incorporates ddNTPs, each labeled with a specific fluorescent dye. The successive cycles of denaturation, annealing and extensions during the sequencing reaction will produce linear amplifications of template. By incorporating specifically dye-labeled ddNTPs the reactions simultaneously terminate and label each product with a particular ddNTP. When the reaction mixtures

T7 Promoter Primer: 5' TAATACGACTCACTATAGGG 3'

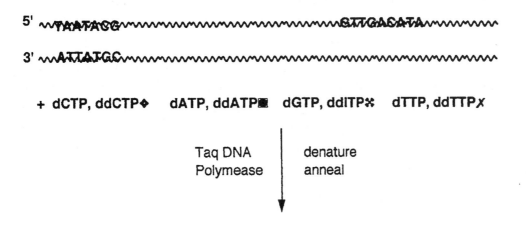

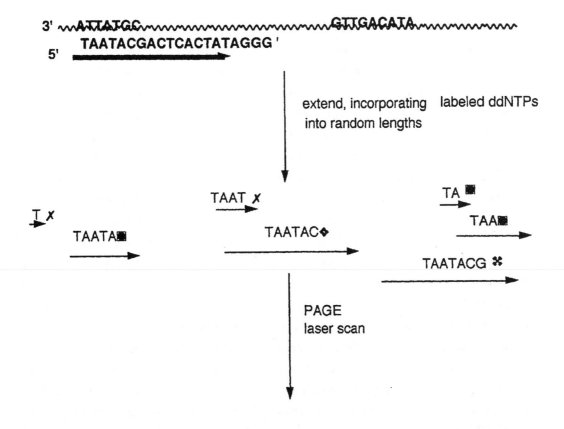

Figure 11.1   ABI Prism™ BigDye™ dideoxy terminator cycle sequencing

Table 11.1*

Terminator	Acceptor dye	Color on electropherogram
A	dR6G	Green
C	dROX	Red
G	dR110	Blue
T	dTAMRA	Yellow

The donor dye molecules shown above are linked to a dichlororhodamine dye (dRhodamine) that serves as an acceptor.

Materials and Methods

Materials

Concentration determination

♦ P-20 Pipetmen and sterile white or yellow tips

♦ Purified plasmid DNA samples from Experiment 10

♦ Calf thymus or λ standard DNA solutions and water blank

♦ Ethidium bromide test solution (1 µg/ml) in water

Ethanol precipitation

♦ P-20 and P-200 Pipetmen and sterile white or yellow tips

♦ Gel-loading or other fine tips for P-200 or a Pasteur pipet drawn out to a fine tip

♦ 7.5 M ammonium acetate (AmOAc)

♦ 95% (v/v) ethanol

♦ 70% (v/v) ethanol

♦ nuclease-free water (autoclaved distilled deionized water)

♦ SpeedVac, if available

Cycle sequencing

♦ Thermal cycler

♦ Sequencing Mixture (BigDye™) prepared by ABI Prism® (Applied Biosystems) version 3.0: ddATP-R6G; ddGTP-R110; ddCTP-ROX; ddTTP-TAMRA; dATP;

*Reproduced by permission of Applied Biosystems.

dITP; dCTP; dTTP; Tris–HCl (pH 9); MgCl$_2$; thermostable pyrophosphatase; AmpliTaq® DNA polymerase

♦ 5× sequencing buffer (50 mM Tris-HCl, 10 mM MgCl$_2$, pH 9)

♦ Nuclease-free deionized water

♦ Mineral oil (preferably autoclaved)

♦ Forward primer (T7 promoter; 1 pmol/μl)

♦ Purified plasmid template DNA (30–50 ng/μl) from Experiment 10

Method

Estimate total plasmid DNA

Estimate the concentration of DNA in a 50 μl sample using the ethidium bromide spot test method, to determine how much water will be needed for resuspending your pellet to obtain the required template concentration.

(1) For a standard curve, make a row of seven 4 μl ethidium bromide test solution drops on a piece of plastic film placed on a transilluminator.

(2) Place a 4 μl drop of ethidium bromide test solution below the row.

(3) Add 4 μl of DNA standard to each drop in the standard curve. Mix by pipetting up and down once. Use a fresh tip for each concentration.

(4) Using a fresh tip, add 4 μl of a resuspended template DNA from the previous experiment to the sample drop. Mix once.

(5) Put on goggles, turn on the transilluminator and compare the sample to the standard curve. Estimate the concentration of the sample.

(6) Calculate how much total DNA is in the sample, then determine how much nuclease-free water will be required to resuspend the precipitated DNA to obtain at least 300–500 ng of template DNA in a final volume of 10–20 μl (at least 30–50 ng/μl, more if possible).

(7) Remove a 2–5 μl aliquot to be stored in the freezer for transforming the expression host after the sequence has been verified. If all the DNA must be used for the sequencing, additional plasmid will have to be grown up and purified for transformation.

Ethanol precipitation

Room temperature precipitation, aided by ammonium acetate is used here. There is fine line between getting plasmid DNA precipitated and not precipitating any nucleotides or other contaminating materials from the plasmid preparation.

(1) Add 25 μl 7.5 M ammonium acetate to the remaining plasmid DNA.

(2) Add 200 μl 95% ethanol to the mixture.

(3) Place the tube in a microcentrifuge, hinge facing out. Centrifuge at $10\,000–14\,000\,g$ for 20 min to pellet the DNA.

(4) Using a fine-drawn Pasteur pipet or a P-200 Pipetman fitted with a gel-loading tip (fine tip) slide the tip down the side of the tube away from the pellet. Note that the pellet cannot be seen. If the tube was centrifuged hinge out then the pellet is on the hinge side. Very carefully aspirate the supernatant.

(5) Wash the pellet by adding 250 μl 70% ethanol to the tube without disturbing the pellet.

(6) Centrifuge as in step (3).

(7) Aspirate the ethanol as in step (4).

(8) Carefully air dry the pellet or dry it in a stream of nitrogen, or in a SpeedVac, if one is available. The pellet can also be dried in a heat block at 90°C for 1 min. Leave the cap open on the tube.

(9) Add enough nuclease-free water to obtain an estimated concentration of 20–50 ng/μl in 15 μl or less. Up to 10 μl containing 200–500 ng can be used for for the sequencing and 4–5 μl for a concentration estimate.

(10) To be certain you have enough DNA, estimate concentration using a fresh ethidium bromide standard curve as described above. Freeze any remaining sample. To obtain a more exact estimate, 4–5 μl DNA can be run on a thin agarose gel and the amount estimated in ng using mass markers in one lane.

Sequencing reaction

(1) Set up the reaction in a fresh PCR tube then vortex briefly to mix:
4 μl BigDye™ sequencing reaction mixture

are electrophoresed, the terminated PCR products are separated by size. A laser scan is used to detect the fluorescent dye corresponding to the specific ddNTP that terminated the chain fragment. Figure 11.1 shows graphically how the process works. The sequence of fluorescent dyes will correspond to the DNA sequence of the DNA template. The forward primer (T7 promoter primer) chosen is $5' \rightarrow 3'$, and immediately precedes the LTB target gene. It anneals (is complementary to) to the $3' \rightarrow 5'$ strand of the template DNA and, when extended, will read through the T7 promoter sequence and the LTB target gene if enough time is allowed for the extension step to complete the LTB sequence. The T7 promoter primer is used here to allow verification that the inserted LTB gene is properly oriented in the plasmid with respect to the promoter site and can therefore be transcribed correctly. Any false stops, caused by the polymerase enzyme prematurely falling off the template DNA, are not visualized because no dye-labeled ddNTP has been attached. A reverse primer is not used because sequential extensions from the forward primer are read.

Sequencing reaction mixture

A prepared reaction mixture called BigDye^{TM} will be used. The mixture includes the dye-labeled terminators, deoxy nucleoside triphosphates, AmpliTaq® DNA polymerase (trade name for the *Taq* polymerase supplied by ABI), a thermostable recombinant pyrophosphatase, magnesium chloride and buffer. *Taq* (AmpliTaq®) DNA polymerase works satisfactorily for cycle sequencing, but requires very high concentrations of the dye-labeled ddNTPs, since it does not bind these modified nucleotides very well. For this reason, the completed reactions contain a large quantity of unincorporated dye-labeled terminators as well as the labeled extension products. The unincorporated terminators are removed by precipitating the labeled extension products from solution and discarding the supernatant, a procedure called 'clean-up', or purifying extension products.

In the BigDye^{TM} reaction mixture, buffer, *Taq* polymerase concentration and the concentration of reactants have been optimized to give a balanced distribution of signal for base-pair extensions that include base 10 to base 700 from the primer. AmpliTaq® DNA polymerase enzyme is a variant of the *Thermus aquaticus* DNA polymerase that contains a point mutation in the active site. This results in less discrimination against dideoxy nucleotides, easing their incorporation into the PCR products. In addition the $5' \rightarrow 3'$ nuclease activity has been removed by mutation and it is formulated with a thermally stable inorganic pyrophosphatase so that phospholysis at higher temperatures is not a problem. The dNTP mix substitutes dITP (deoxyinosine triphosphate) in place of dGTP to minimize band compression (crowding of bands). The gel will be scanned by an argon ion laser. The donor dyes are rhodamine derivatives:

2 μl 5× sequencing buffer*
2–10 μl (200–500 ng) pET28LTB template DNA
3.2 μl T7 promoter primer (1 pmol/μl = 3.2 pmol total) and
nuclease-free water to 20 μl final volume.

(2) Add 20 μl mineral oil to the top of the liquid to prevent any possible evaporation, even when using a thermal cycler with a heated lid.

(3) Run (see Applied Biosystems Protocol booklet) for a GeneAmp Thermal Cycler:
96°C for 10 s
50°C for 5 s
60°C for 4 min
Repeat for 24 more cycles
Hold at 4°C. It can be held overnight if necessary. For all other thermal cyclers, run: 96°C for 30 s; 50°C for 15 s; 60°C for 4 min. Repeat for 24 more cycles. Hold at 4°C. It can be held overnight if necessary.

(4) When reaction is complete spin tubes briefly and freeze.

Preparation for experiment

(1) DNA standards (see Experiment 3).

(2) Ethidium bromide test solution (see Experiment 3). To obtain a more exact estimate 4–5 μl DNA can be run on a thin agarose gel and the amount estimated in ng using mass markers (for example, Mass Ruler, BioRad catalog no. 170-8207 or 170-8356) in one lane.

(3) BigDye™ is commercially available from Applied Biosystems (formerly PE Biosystems), Foster City, CA, USA.

(4) 5× sequencing buffer: 500 mM Tris–HCl/10 mM $MgCl_2$ pH 9.0. Make Trizma Pre-Set Crystals pH 9 (Sigma catalog no. T6003) to 500 mM and add the requisite amount of $MgCl_2$.

References

Sambrook, J. and Russell, D.W. (2000). *Molecular Cloning: a Laboratory Manual*. Cold Spring Harbor, NY: Cold Spring Harbor Laboratory Press.

---

*The BigDye™ is halved in this protocol for economy. The extra buffer is added to compensate for the dilution of buffer components in the BigDye™ reaction mixture.

EXPERIMENT *12*

(a) Purification of Extension Reaction Products
(b) Sequencing Gel Demonstration

Introduction

Reaction mixture purification

The extension reaction mixture obtained in the previous experiment must be manipulated to remove all unincorporated dye-tagged molecules and enzymes, leaving only extension products to be run on a sequencing gel. Clean samples provide clean unambiguous gel results. The extension products can be precipitated by several methods. Room temperature isopropanol is used here because it is a convenient, uncomplicated and effective method of purification. It is done at room temperature because one must precipitate extension products without precipitating unincorporated dye-labeled nucleotides. The oil added to prevent evaporation must be removed from the reaction mixture. This can be accomplished by centrifuging the mixture to separate the mineral oil from the aqueous fraction, then carefully pipetting the aqueous fraction out of the reaction tube. Alternatively, it can be removed by relying on the attraction of one hydrophobic substance for another. The entire mixture, oil and all, is pipetted onto a piece of Parafilm. Parafilm is a hydrophobic wax film and, upon rolling the drop around on the film, one finds that the mineral oil adsorbs to the hydrophobic film and the aqueous drop can be separated from the oil as a tight, spherical hydrophilic droplet. The aqueous drop is then transferred to a 1.5 ml microcentrifuge tube for further purification.

Isopropanol precipitation times shorter than 15 min will result in loss of very short extension products. Precipitation times longer than 24 h will increase the precipitation of unincorporated dye terminators (see Applied Biosystems Protocol, 'ABI Prism$^{®}$ BigDye^{TM} Terminator', version 3.0; Ready Reaction Cycle sequencing kit). Ethanol precipitation may leave traces of unincorporated terminators, which will be seen at the beginning of the sequence data up to base 40, and some loss in recovery may occur, but this method is the current practice, although there appears to be no discernible difference between the results of isopropanol and ethanol precipitation. Both protocols are given here.

Sequencing gels

Agarose gels do not have the fine resolution necessary to separate DNA fragments having very close molecular weights, for example those differing by only one base pair. Polyacrylamide gel electrophoresis (PAGE) is the method of choice for separating DNA fragments of very closely related size. The gels used are made from polyacrylamide. Acrylamide ($H_2C = CH - C\,O - NH_2$) is formed into long chains by vinyl polymerization. The acrylamide monomer is a potent neurotoxin, causing numbness due to blockage of nerve transmission when it binds to nerve endings. The chains are cross-linked into a mesh-like structure by polymerization in the presence of a bi-functional co-monomer. The most common cross-linking agent is N,N'-methylene-bis-acrylamide (called 'Bis') $CH_2 = CH - CO - NH - CH_2 - NH - CO - CH = CH_2$. The polymerization reaction produces random chains of polyacrylamide incorporating a small proportion of Bis molecules, and these can then react with groups in other chains to form cross-links in a three-dimensional network. Polyacrylamide is not toxic. The concentration of acrylamide monomer determines the average polymer chain length while the Bis determines the extent of cross-link formation or mesh size. Gel density, elasticity, mechanical strength and pore size are determined by the relative concentrations of each component. The polymerization of monomers proceeds via a free-radical mechanism initiated by the addition of ammonium persulfate, which produces free oxygen radicals in the presence of the catalyst N,N,N',N',-tetramethylenediamine (TEMED). It is also possible to initiate polymerization photochemically by the addition of riboflavin, which is photo-decomposed to leucoflavin. In the presence of traces of oxygen, re-oxidation of leucoflavin occurs with free-radical generation, leading to gelation (polymerization) of the acrylamide–Bis mixture (Andrews, 1986).

The gel is cast between two glass plates (with combs as well-formers) as soon as the catalyst is added to the monomer/persulfate mixture; then it is allowed to polymerize.

Gel size is determined by the size of the fragment to be sequenced, since there should be one band (terminated fragment) for each base pair. Very long gels are therefore required.

Another method, capillary electrophoresis, is also used for sequencing. In this technique, an instrument from Applied Biosystems (ABI) uses glass capillary tubes. Each sample runs separately through a capillary tube that is filled automatically by a syringe with a material that ABI calls a 'performance optimized polymer'. The electrophoresis occurs through the polymer in the capillary, rather than through a gel. Very small amounts of material are required and the procedure is very rapid, with very high throughput. Techniques of this type have greatly increased the number of genomes sequenced in recent years. More information can be found at the ABI website (www.appliedbiosystems.com).

Analysis of the bands can be complicated by compressions, clustering of bands, that occur when two or more single-stranded DNAs of different lengths migrate through a polyacrylamide gel with the same mobility. The compressions occur when the DNA is not fully denatured during electrophoresis. This results in anomalous spacing and consequent difficulty reading the sequence. They usually affect G and C tracks, although A and T tracks may also be affected. Details of this problem and its solution can be found in Sambrook and Russell (2000; Vol. 2, p. 12.109).

Materials and Methods

Materials: reaction mixture purification

♦ Microcentrifuge

♦ Heat-block or thermal cycler (90°C) or a SpeedVac, if available

♦ P-20 and P-200 Pipetmen, sterile white or yellow tips

♦ Parafilm

♦ Gel-loading tips or a drawn-out Pasteur pipet

♦ Sterile 1.5 ml microcentrifuge tubes

♦ 75% (v/v) isopropanol

♦ 100% isopropanol

For alternative ethanol precipitation:

♦   3 M sodium acetate (NaOAc) pH 4.6

♦   non-denatured 95% ethanol

♦   deionized water

♦   Reaction mixture from Experiment 11

Method

(1) Thaw the reaction mixture from Experiment 11. Spin briefly to collect contents.

(2) Using a P-200 Pipetman, slide the tip to the bottom of the tube. Holding the tube at eye level, carefully remove the lower, aqueous layer. Transfer as little of the oil as possible.

(3) Dispense the drop onto a piece of Parafilm. Using the pipet tip, roll the droplet away from the adhering oil. Roll and pull the drop away from the slick once or twice to remove all traces of oil.

(4) Pick up the aqueous drop with a fresh tip and dispense it into a clean 1.5 ml tube.

(5) Add 80 ml of 75% isopropanol (or 20 μl deionized water and 60 ml 100% isopropanol). Final isopropanol should be $60 \pm 5\%$. *Isopropanol can be harmful if inhaled or ingested. It can cause central nervous system depression and be irritating to the eyes. Be careful.*

(6) Close the tube and vortex briefly.

(7) Leave the tube at room temperature for 15 min to precipitate the extension products without precipitating the labeled nucleotides.

(8) Place the tube in a microcentrifuge, hinge side facing out. Centrifuge 10 000–14 000 g for 20 min.

(9) Immediately aspirate the supernatant with a fine-drawn Pasteur pipet or a gel-loading tip. Use a separate pipet tip for each sample. The pellet will be invisible, so slide the tip in, on the opposite side of the tube, and carefully withdraw the supernatant. The supernatant must be removed completely since it contains unincorporated dye terminators dissolved in it. These will compromise the gel data.

(10) Wash the pellet by adding 250 μl 75% isopropanol to the tubes. Vortex briefly.

(11) Place the tubes in the microcentrifuge in exactly the same orientation as in step (8). Spin for 5 min at maximum speed.

(12) Aspirate the supernatant carefully, as described in step (9).

(13) Dry the samples in a vacuum centrifuge (SpeedVac), if one is available, for 10–15 min, or place the tubes, lids open, in a heat block or thermal cycler at 90°C for 1 min.

The samples are now ready to be dissolved in loading buffer and loaded onto a sequencing gel. As an alternative, Applied Biosystems recommends an ethanol/ammonium acetate precipitation for use with BigDye terminators version 3.0, as follows:

(1) Remove the oil from the reaction mixture as described in steps (2)–(4) above.

(2) In a separate tube prepare (final volume 80 μl): 3.0 μl 3 M sodium acetate (NaOAc) pH 4.6; 62.5 μl non-denatured 95% ethanol; 14.5 ml deionized water.

(3) Add the prepared 80 μl mixture to the 20 μl of reaction mixture.

(4) Close the tubes and vortex briefly.

(5) Leave the tubes at room temperature for 15 min to precipitate the extension products, leaving the unincorporated nucleotides and dye terminators in solution.

(6) Place the tubes in a microcentrifuge and mark their orientations (hinge in or hinge out).

(7) Spin for 20 min at maximum speed.

(8) Carefully aspirate the supernatant. The pellet may not be visible.

(9) Add 250 ml of 70% (v/v) ethanol to the tubes and mix briefly.

(10) Place the tubes in the microcentrifuge in the same orientation as before.

(11) Spin at maximum speed for 5 min.

(12) Aspirate the supernatants carefully.

(13) Air dry or dry in a SpeedVac, if one is available.

If there is one available, students should have an on-site visit to see the sequencing instrument.

References

Andrews, A. T. (1986). *Electrophoresis, Theory, Techniques, and Biochemical and Clinical Applications*, p. 482. Oxford: Oxford University Press.

Sambrook, J. and Russell, D.W. (2000). *Molecular Cloning: a Laboratory Manual*. Cold Spring Harbor, NY: Cold Spring Harbor Laboratory Press.

EXPERIMENT *13*

(a) Analysis of Sequencing Data
(b) Verification of Insert Sequence

Introduction

Bioinformatics

Bioinformatics is the analysis of DNA sequences and the interpretation of sequence data, thus it is the generation of new knowledge from existing data. This type of research includes the development and testing of software tools necessary to generate new knowledge from primary source information deposited in databases and the literature. There are several important computer programs available for such analyses. These include BLAST (Basic Local Alignment Search Tool), which performs sequence alignment and GCG, a computer program that analyzes DNA and protein sequences and makes structural predictions based upon the sequences. BLAST was developed by Altschul *et al.* (1997). It is accessed from the public domain website, www.ncbi.nlm.nih.gov/blast, produced by the National Center for Biotechnology Information (NCBI), part of the National Library of Medicine (NLM), at the National Institutes of Health (NIH, Bethesda, MD).

'BLAST is a set of similarity search programs designed to explore all of the available sequence databases regardless of whether the query is protein or DNA. The BLAST programs have been designed for speed, with a minimal sacrifice of sensitivity to distant sequence relationships. The scores assigned in a BLAST search have a well-defined statistical interpretation, making real matches easier to distinguish from random background hits. BLAST uses an heuristic algorithm which seeks local as opposed to global alignments and is therefore able to detect relationships among sequences which share only isolated regions of similarity' (Altschul *et al.*, 1997). For a better understanding of BLAST you can refer to the BLAST Course (www.ncbi.nlm.nih.gov/blast), which explains the basics of the BLAST algorithm.

BLAST subroutines

Blastn searches the databases for nucleotide matches, blastp searches for amino acid sequence matches, blastx searches for distant protein homologs in a sequence database, tblastn searches a nucleotide database translated in all six frames using a protein sequence query, and tblastx searches the six-frame sequence database translations of a nucleotide sequence database using six-frame translations of a nucleotide query sequence. A nucleotide query sequence can be input using a FASTA format, in which the first line can be 'letter and numerical description', while the second line and following lines are the one-letter nucleotide sequence, written continuously with no punctuation. One can simply copy the sequence from a source and paste it into the box for this purpose, or type in the sequence directly. Alternatively, the databank accession number can be given. In that case, the program will search the databank for the data associated with that number and use the information to search for matches.

Nucleotide and amino acid sequences are deposited by investigators into several databases, all of which are accessible by the public. These include the Protein Data Bank (PDB), GeneBank, Swiss-Prot, and both the NIH and TIGR human genome data banks, as well as drosophila, mouse and *E. coli* databanks. A few examples, with their descriptions, are shown below. For a more complete description and a list of all available databases go to www.bionavigator.com

PDB is the single international repository for the processing and distribution of three-dimensional macromolecular structure data primarily determined experimentally by X-ray crystallography and NMR. PDB is operated by the Research Collaboratory for Structural Bioinformatics (RCSB) under contract to the US National Science Foundation and is supported by funds from the National Science Foundation, the Department of Energy, and two units of the National Institutes of Health: the National Institute of General Medical Sciences and the National Library Of Medicine.

GenBank is part of the International Nucleotide Sequence Database Collaboration, which is a collaboration between the DNA DataBank of Japan (DDBJ), the European Molecular Biology Laboratory (EMBL), and GenBank at NCBI. These three organizations exchange data on a daily basis.

SWISS-PROT is a database maintained by the Department of Medical Biochemistry of the University of Geneva in collaboration with the EMBL Outstation, EBI (European Bioinformatics Institute) in the UK. It is the most extensively annotated of all the sequence databases and is very useful for this reason. Data is derived primarily from the translation of DNA sequences in the EMBL nucleotide sequence database. Other sequences are obtained from PIR, extracted from literature, or submitted directly by researchers. Data from these sources is substantially annotated before being included.

Prosite is a database of protein families and domains. Protein signatures are used to assign newly sequenced proteins to a specific family of proteins and thus to formulate hypotheses about its function. PROSITE currently contains signatures specific for about a thousand protein families or domains. Each of these signatures comes with documentation providing background information on the structure and function of these proteins.

Other resources available from NCBI

The on-line BLAST Course was written by Dr Stephen Altschul and discusses the basics of the Gapped BLAST algorithm. In addition the full text of the 1997 Nucleic Acids Research paper 'Gapped BLAST and PSI-BLAST: a new generation of protein database search programs' is also available on-line at www.ncbi.nlm.nih.gov/blast. Chapter 7 of the *Cold Spring Harbor Genome Analysis Laboratory Manual* also provides helpful introductory information for users of molecular biology databases and software. This chapter is available over the world wide web from the Cold Spring Harbor Laboratory home page (www.cshl.org/) under Cold Spring Harbor Laboratory Press (www.cshlpress.com).

There are many sites which offer software tools for molecular biologists and for manipulating sequence data. Some of the largest of these are listed below:

European Bioinformatics Institute (EBI) BioCatalog, www.ebi.ac.uk/biocat

Indiana University IUBio Archive, iubio.bio.indiana.edu

Pedro's BioMolecular Research Tools, www.public.iastate.edu/~pedro/research_tools.html

Sample analysis

Data from samples analyzed on an ABI Prism Automated DNA Analyzer (Applied Biosystems) should be available on disk. The free program Chromas allows graphic visualization of the peaks and lists the observed sequence. The specifically dye-labeled ddNTPs were incorporated into terminated chains, and the reaction mixture electrophoresed. Each of the four colored curves represents the electrophoretic pattern of fragments containing one of the dideoxy nucleotides, for example, green, red, black and blue correspond to ddATP, ddTTP, ddGTP and ddCTP, respectively. The 3′-terminal base of each oligonucleotide fragment is identified by the laser-excited fluorescence of its gel band. A detailed discussion can be found in Voet *et al.* (1999).

Transformation

The transformation protocol from Experiment 8 will be followed, but this time using an expression host rather than a cloning host. The difference is that the expression host carries a plasmid (λDE3 lysogen) that contains the nucleotide sequence for the T7 RNA polymerase gene, plus several control elements for turning on and off the expression of the T7 RNA polymerase protein. The RNA polymerase protein binds the T7 promoter DNA on the target gene construct [which will be introduced into the BL21(λDE3 lysogen) host bacteria]. Upon binding the promoter sequence, the polymerase catalyzes the transcription of mRNA corresponding to the target gene. The expression control strategy is described in detail in Experiment 14.

Materials and Methods

Materials

Computer analyses

♦ Mac or PC with Internet connection and Microsoft Word

♦ Printer (optional)

♦ Chromas visualization program for data display (PC) or ABI Edit-View (for Mac)

♦ Sequence data

♦ A file conntaining the A2B nucleotide sequence in one-letter code as shown in Experiment 3

♦ A file containing the LTB amino acid sequence in one-letter code

Method: simple alignment analysis using BLAST

This section can be done in class or done prior to class, but must precede transformation. The transformation protocol takes about 90 min.

Nucleotide alignment

(1) Using Netscape (or Explore or other web browser) go to www.ncbi.nlm.nih.gov/blast/.

(2) Click on 'BLAST Info/Tutorial' (on the left sidebar) and click on 'What is Similarity Searching' and then on 'Glossary of Terms' (on left sidebar).

(3) Return to BLAST Info/Tutorial and choose 'Query Tutorial'. Read through the tutorial.

(4) Now go back to the BLAST homepage to 'Nucleotide BLAST'. Choose standard nucleotide–nucleotide BLAST blastn. Use database PDB to limit results.

Do a practice run with a known sequence:

(5) Open the file A2LTB/FASTA format on the Desktop, highlight and copy the A2LTB sequence (sequence only). It is the original Dallas and Falkow (1979) sequence for the *E. coli* gene *elt*B of porcine (pig) origin (see Experiment 3).

(6) The sequence is already in FASTA format; it can be copied from the file and pasted directly into the indicated box.

(7) Do a search by clicking on the 'BLAST!' button.

(8) Click on 'Format' to get the results. Scroll down past the bar graph to the 'hit list'.

(9) Are there any hits? If not, return to the search screen and change the database line to 'nr' then search again.

(10) Examine the first two hits (98%) and identify any nucleotide mismatches. Find the MET (atg) and the Ala#1 (gct) and the STOP (tag) codons. Answer the following questions:
Are any of the mismatches within the gene?
Are there any that will result an amino acid substitution?
Examine several of the other sequences shown. Are the base substitutions found in those sequences? Which sequence do you suspect is incorrect?
When the nucleotide sequence is transcribed and subsequently translated into an amino acid sequence, will there be a mutant protein or will the PCR of the target gene produce a matched sequence and an unmutated protein?

Does a single nucleotide change necessarily result in a different amino acid sequence in the protein?

If the amino acid sequence is altered at a single position in the protein, at which positions in the protein might mutations be most influential with respect to conformation and quarternary structure LTB protein?

If the amino acid sequence is altered at a single position in the protein, at which positions in the protein might mutations be least influential with respect to conformation and quarternary structure LTB?

What is the function of the LTB protein?

Would an inability to form the proper quarternary structure affect its function?

What about conservative (hydrophobic for hydrophobic amino acids, for example) vs non-conservative mutations (hydrophobic for hydrophilic amino acids or basic for acidic amino acids, as examples)?

Amino acid sequence

(11) Return to the homepage and, in Protein BLAST choose standard protein–protein blastp. Paste the amino acid sequence copied from the LT amino acid sequence file into the FASTA format box. Search by clicking on 'BLAST!' then on 'Format'. This is a practice run with a known sequence. How much homology does the LTBp sequence from Experiment 2 have to cholera enterotoxin B-subunit (>qb|AAD51360.1|AF175708)?

(12) Write out the answers to bulleted questions.

Identity of an unknown sample

If sequence data is available for some other protein, the data can be examined as a test, to determine the identity of the sequenced material.

Method: sequence analysis of sample

The sample nucleotide sequence has been downloaded onto a disk. It can be visualized using an Edit-View program called Chromas. To view the sequence data on a PC:

(1) Open Chromas (on the desktop).

(2) In the File menu, open the sequence file (*.seq) where * indicates the operator ID.

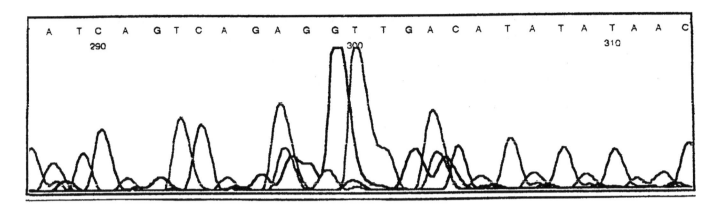

Figure 13.1 Portion of a typical electropherogram data from an ABI Prism® sequencer

(3) Observe that the nucleotide sequence is displayed above the fluorescence peaks. Note that the letter N in the sequence means that the laser identified a band in that position but could not determine the specific base. Print the file.

(4) Click on the sequence text line. Then, in the menu, go to 'Edit > copy > FASTA format' to copy the sequence.

(5) Close the Chromas file, keep the Clipboard, go to File and open a new document. Paste the copied sequence into the new document. Close and save the new document. Keep the Clipboard.

(6) Open Netscape Navigator (or Explorer) from the desktop. Go to www.ncbi.nlm. nih.gov/blast.

(7) Find Nucleotide BLAST.

(8) Choose standard nucleotide–nucleotide blastn for the program, choose 'nr' for the database. Paste the FASTA formatted sequence from the Clipboard into the indicated input box. Click on 'BLAST!'; then click on 'Format results'.

(9) When the results appear, scroll to the colored 'hit table', then to the sequence alignments. Check your query against the sequence with the highest percentage homology. Your sequence is from a porcine strain of *E. coli*. Note that Ns are not matched. Print the first two or three pages of the search results. Scan through the rest noting what other genes have significant sequence homology.

(10) Go to the printout and see if you can identify the Ns by comparing them to the known sequence and the fluorescence peaks. Fill in as many as possible.

(11) If there is output from the T7 promoter primer you can use tblastn to check that the insert is in frame. Find the sequence for the T7 promoter forward primer; see

Experiments 9 and 11. Check the primer sequence against the sequence alignment. *Hint*: highlight, copy and paste the BLAST nucleotide sequence alignment into a Microsoft Word document. Use the 'Find' function to search for the Forward primer sequence (exclude the *Nco*I site at the 5′ end of the primer).

(12) If the sample sequence matches the *E. coli* enterotoxin (porcine) sequence, the sample insert is correct. Proceed to Transformation of Expression Host (Experiment 14). In the event that the pET28LTB sequence cannot be determined or is grossly incorrect, transformation of the expression host can proceed with purified pTZLT18 or pTZΔSA2B plasmids. Protocols should be analyzed to determine the source of error.

Preparation for experiment

(1) File A2LTB/FASTA format (see Experiment 2, LTp sequence). Prepare a file named A2B.FASTA from the one-letter code nucleotide sequence shown in Experiment 3. Prepare a file named LTBaa.FASTA using the one-letter amino acid code from the amino acid sequence in Experiment 3. Files can be downloaded from www.chemistry.montana.edu/~bspangler/methods.html

(2) ABI Prism Edit View for Mac is the ABI automated sequence viewer application. It can be downloaded from www.applied biosystems.com (software section). Chromas is the PC version for viewing ABI sequencing data. It is freeware obtained at www.technelysium.com.au/chromas.html – choose V1.45 from the sidebar.

References

Altschul, S. F., Madden, T. L., Schaffer, A. A., Zhang, J., Miller, W. and Lipman, D. J. (1997). Gapped BLAST and PSI-BLAST: a new generation of protein database search programs. *Nucleic Acids Res.* **25**, 3389–3402.
Voet, D. J. G., Voet, J. G. and Pratt, C. W. (2000). *Fundamentals of Biochemistry*, pp. 59–63. New York: John Wiley.

EXPERIMENT 14

(a) Gene Expression
(b) Transformation into an Expression Host

Introduction

The phrase 'gene expression' means that the gene is relayed from the DNA to an amino acid sequence (protein). For this to happen DNA encoding a particular protein is transcribed by DNA-dependent RNA polymerase, which catalyzes the formation of mRNA by arranging the complementary ribose nucleotides then catalyzing phospho-diester bond formation. The mRNA, together with tRNA associated with specific amino acids, and ribosomal RNA, which catalyzes amide bond formation, translates the nucleotide sequence encoded in mRNA into an amino acid sequence (a protein or peptide).

DNA plasmid vector pET28b from Novagen Inc. has been used to make the DNA construct, pET28LTB. Details of the system can be found at www. novagen.com. The pET 28b sequence is shown in Figure 5.1. The plasmid pET28b contains, among other sequences, a T7 promoter sequence, a T7 terminator sequence, a section into which an insert may be positioned between the promoter and terminator sequences, and a Kanamycin resistance gene. The T7 promoter is a bacteriophage promoter sequence not recognized by host cell *E. coli* RNA polymerase and, therefore, virtually no protein expression encoded by the plasmid occurs during host bacterial growth. The genes under control of the introduced plasmid construct pET28LTB cannot be expressed by the bacterial host *E. coli* BL21(λDE3) until an inducer is supplied which allows expression of T7 RNA polymerase.

The strategy is as follows: once established in a cloning host such as *E. coli* BL21, the plasmid is isolated then transferred to an expression host, here *E. coli* BL21(λDE3). The expression host bacterium contains a chromosomal copy of the RNA polymerase gene from bacteriophage T7 under control of *lac* UV5, a regulatory site in the DNA sequence. Bacteriophage T7 is a λ phage (Studier and Moffatt, 1986). Bacteriophages (bacterial viruses) are small DNAs that infect bacteria. Expression of the T7 RNA polymerase gene from the phage DNA incorporated into the *E. coli* BL21 chromosome is repressed by the binding of *lac* repressor protein, which binds to the control (regulatory) nucleotide sequence. In addition, the T7 promoter that controls expression of the LTB insert is interrupted by the presence of the *lac* operator sequence, a regulatory site that prevents RNA polymerase binding and subsequent mRNA synthesis when it is bound and occluded by *lac* repressor protein. When the *lac* operator sequences that control T7 polymerase expression as well as the T7 promoter region controlling LTB expression are bound by *lac* repressor protein, expression of the LTB insert is turned off directly. To allow expression of both the T7 polymerase and LTB genes, a compound that binds to *lac* repressor protein is added to the growing cells. The function of this compound is to remove *lac* repressor protein from the operator site (de-repression), thereby permitting expression of the T7 RNA polymerase, which will then be able to transcribe the LTB gene sequence.

The de-repression compound, IPTG is an analog of lactose. Lactose is the natural de-repressor since addition of lactose to the growing cells removes *lac* repressor protein from the operator nucleotide sequence. The polymerase is now able to bind the promoter nucleotide sequence and a β-galactosidase gene normally associated with it can be expressed. IPTG and lactose are called 'inducers' because, by removing the repressor protein, they induce expression of the β-galactosidase gene. By replacing the β-galactosidase gene with the LTB insert this very effective system is turned to advantage. Induction occurs because the IPTG, like lactose, can bind the repressor protein and change its conformation, so that the repressor releases from the operator regions and clears the way for *E. coli* polymerase to bind and transcribe the T7 RNA polymerase and then for the T7 RNA polymerase to bind the T7 promoter and transcribe the LTB gene insert. Thus, addition of IPTG induces T7 RNA polymerase transcription and translation by the host bacterium. The newly expressed polymerase binds to the T7 promoter region on the pET28LTB plasmid DNA in the host cell where it transcribes LTB gene insert into mRNA, which is subsequently translated into LTB protein by host *E. coli* ribosomes. The system is so active that the desired product, LTB, can comprise more than 50% of the total cell protein a few hours after induction. It is illustrated in Figures 14.1 and 14.2.

Two types of host cell are possible: the cloning host and the expression host. The pET plasmid could have been cloned into one of several cloning host derivatives of *E. coli*

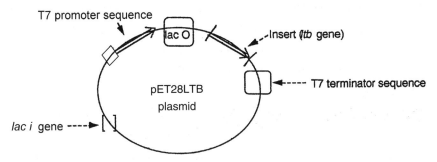

Figure 14.1   Control elements in pET system, where *lac O* is the operator sequence (a control element, site of *lac* repressor protein binding) and the *lac i* gene encodes *lac* repressor protein. T7 promoter is the site where T7 RNA polymerase binds to begin transcription and the T7 terminator sequence causes the polymerase to stop synthesizing mRNA (reproduced by permission of Novagen Inc.)

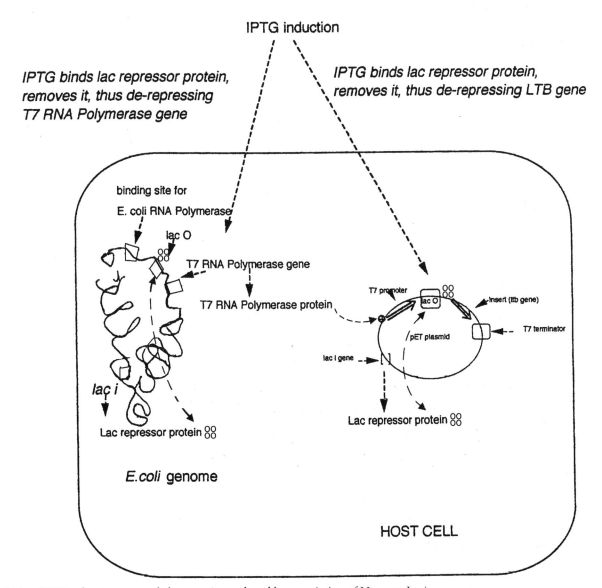

Figure 14.2   IPTG induction: control elements (reproduced by permission of Novagen Inc.)

strain K12. Those bacterial strains, as with other cloning host strains, lack the T7 RNA polymerase gene. The notation for a cloning host would be 'E. *coli* BL21'. These strains lack genes encoding specific proteases that could degrade the target protein. The designation for an expression host, 'E. *coli* BL21(λDE3)' indicates that the expression host contains bacteriophage λ DE3, which carries genes for protein expression. The BL21 strain and its derivatives are probably the most widely used for target gene expression. They were derived from E. *coli* DH5α, itself a derivative of E. *coli* strain K12. Strain K12 is a mutant that is incapable of growing in the natural environment because it has a specific amino acid requirement. The mutation was produced so that, should recombinant research bacteria escape to the environment, they could not survive.

Materials and Methods

Materials

Transformation

♦ Thermal cycler, water bath or heat block with water in block wells set at 42°C

♦ P-20 and P-200 Pipetmen and sterile white or yellow tips

♦ Ice/ice bucket and individual containers

♦ Competent BL21(DE3) (20 µl aliquots in sterile 0.5 or 1.5 ml tubes)

♦ pET28LTB or pTZLT18R plasmid containing insert with correct sequence (1–5 µl), saved in Experiment 11

♦ Test plasmid, supplied with competent cells, if purchased (Novagen; optional)

♦ SOC media (1 ml/team), supplied with purchased cells or prepared

If no plasmid was stored in Experiment 11, a 3–5 ml broth culture of BL21 cells from a sequenced positive colony or from the glycerol stock of a sequenced plasmid from BL21 saved in Experiment 10 should be grown. A plasmid miniprep must then done to obtain DNA for transformation. If no recombinant construct is available, pTZLT18R can be substituted. See note in Preparation for Experiment.

Plating

♦ Glass hockey sticks (spreaders) or Lazy-L presterilized spreaders (Sigma)

♦ 95% ethanol (if glass is used)

◆ Bunsen burners (if glass is used)

◆ 37°C incubator

◆ LB-Kan30 plates

Method

Transformation into expression host E. coli BL21 (DE3)

(1) Aliquots of 20 µl competent cells in a microcentrifuge tube have been prepared. *Keep them cold, on ice.*

(2) (a) Add 1 µl of the pET28LTB plasmid to the ice-cold cells and gently stir with the pipet tip to mix. Do not be rough – the cells are very fragile!

   (b) As a control, to judge transformation efficiency, add 1 µl of test plasmid to another tube containing 20 µl iced cells. This should be done by one or more teams.

(3) Store the cell mixtures on ice for 5 min. Preheat the thermal cycler, water bath or heat block to 42°C.

(4) Wipe the bottom of the tubes. Place them in the thermal cycler set at 42°C (shut the lid) or water bath or heat block containing water in the wells, set at 42°C, for *exactly* 30 s to heat-shock the cells. Dip only the lower portion of the tubes in the water bath.

(5) Remove the tubes and immediately place them in ice. Store them for 2 min.

(6) Add 80 µl room-temperature SOC to each tube. Store the tubes on ice until ready to incubate [step (7)]. SOC medium contains 20 g/l tryptone, 5 g/l yeast extract, 0.5 g/l NaCl and 20 mM glucose. It allows the cells to recover from the shock of transformation in a comfortable growth environment.

(7) When using *E. coli* BL21 (DE3) or most other strains, the mixture must be incubated at 37°C with shaking (250 rpm) for 1h prior to plating to allow the cells to recover from shock.

(8) Place the plates, inverted, in a 37°C incubator to warm and dry.

Plating on selective media (See Experiment 8 for detailed instructions)

Samples.

Plate out 50 µl of pET28LTB-BL21(DE3) SOC mixture on one plate. Mark the plate with date, initials, amount and ID of plated material (including host cell strain!).

Plate 25 μl of the pET28LTB-BL21(DE3) SOC mixture on another other selective agar plate. Mark the plate with date, initials, amount and ID of plated material (including host cell strain!).

Test plasmid: pipet 20 μl fresh SOC onto the center of a selective agar plate. Add 5 μl of the test plasmid–cell mixture to the center of the SOC puddle. Using the same pipet tip, pick up 5 μl SOC from the edge of the puddle and dispense it into the center of the puddle. Spread the plate as described previously.

(9) Let plates sit 10 min, covered, on the benchtop to allow excess media to soak into the agar.

(10) Invert the plates and label them around the bottom edge with group identification, name of construct, name of host cell and date they were plated.

(11) Place the plates, inverted, in a 37°C incubator.

(12) Check plates after 24 h. If necessary, allow colonies to grow another 24 h before taping edges of plates with Parafilm and storing at 4°C. See Experiment 8 for troubleshooting.

Preparation for experiment

(1) If no plasmid was stored in Experiment 11, a 3–5 ml broth culture of BL21 cells from a sequenced positive colony or from the glycerol stock of a sequenced plasmid from BL21 saved in Experiment 10 should be grown. A plasmid miniprep must then be done to obtain DNA for transformation. See Experiments 9 and 10 for detailed protocols. If no recombinant pET28LTB is available, pTZLT18R (the original template) can be substituted in this experiment. The pTZLT18R plasmid contains the entire LT operon, including the LT holotoxin gene and all regulatory sequences.

(2) Aliquot purchased *E. coli* BL21 (λ DE3) cells (Novagen Inc., Madison, WI, USA, catalog no. 69450-3). Cells should be kept at −70°C. Handle only the very top of the tube and tube cap to prevent warming. To mix the cells, flick the tube, *never vortex*. Cells can be thawed once to aliquot them, then re-frozen at −70°C. Additional freezing and thawing will seriously compromise their ability to be transformed. To aliquot from a 0.2 ml stock, place 10 sterile 0.5 ml tubes (or 0.2 ml PCR tubes) in ice to chill. Remove the 0.2 ml stock from the −70°C freezer and immediately place on ice. To dispense, remove the tube of cells from ice, flick once or twice to mix prior to opening the tube. Remove a 20 μl aliquot from the

middle of the cells and replace the stock tube in ice immediately. Place the aliquot into the bottom of the pre-chilled 0.5 ml tube, mix by pipetting up and down once, then quickly close the tube and replace in ice. After all the aliquots are taken, return any unused tubes to the freezer before proceeding with the transformation.

Alternatively, to prepare competent *E. coli* cells use rapid one-step preparation (Ausubel *et al.*, 1999, pp. 1–28). A longer protocol is also available: prepare an overnight culture of cells in LB broth (see Experiment 5); dilute it 1:100 in fresh LB-broth; grow at 37°C to an OD_{600} of 0.3–0.4; add 1 vol. of ice-cold 2× TSS (see recipe below) and gently mix on ice. Small aliquots may be frozen at −70°C for future use. They should be thawed on ice immediately before use.

2× TSS (transformation and storage medium): dilute sterile (autoclaved) 40% polyethylene glycol (PEG) 3350 to 20% PEG in sterile LB medium containing 100 mM $MgCl_2$. Add dimethylsulfoxide (DMSO) to 10% (v/v) and adjust the pH to 6.5.

(3) SOC medium: 20 g tryptone; 5 g yeast extract; 0.5 g NaCl; 20 mM glucose; distilled water to 1 l. Autoclave to sterilize.

(4) LB-Kan30 plates; see Experiment 5.

This class time lends itself to a comprehensive review of gene expression. A sample lecture outline follows:

(a) Types of RNA – structure and function, mRNA, tRNA and rRNA.

(b) Transcription – DNA-dependent RNA polymerase; operon structure; initiation of transcription; promoter sequences; regulation of transcription; and release of mRNA and binding by ribosome.

(c) Translation – binding of message; synthesis and alignment of tRNA; chemistry of amide bond formation; release; post-translational modification.

(d) T7 bacteriophage – source for pET vectors; PET system (see www.novagen.com); antibiotic resistance; regulatory mechanisms; and host cells.

References

Ausubel, F. M., Brent, R., Kingston, R. E., Moore, D. D., Seidman, J. G., Smith, J. A. and Struhl, K. (1999). *Short Protocols in Molecular Biology*. New York: John Wiley.
Studier, E. W. and Moffatt, B. A. (1986). Use of bacteriophage T7 RNA polymerase to direct selective high-level expression of cloned genes. *J. Mol. Biol.* **189**, 237–248.

EXPERIMENT 15

Induction of LTB Gene Expression: Determination of Time to Maximal Expression

Introduction

Cells for this experiment are grown to an $OD_{600} = 0.6$. This spectrophotometric measurement of the optical density of the cell culture is taken in the visible light range (600 nm) and represents a measurement of turbidity. At the turbidity reading $OD_{600} = 0.6$ the cells are considered to be in the middle of the exponential (cell doubling) phase of the growth curve. Beyond this point, slightly past $OD \approx 1.0$, at the plateau phase of the growth curve, cells are dying at a rate equal to the rate of cell division. Precipitous cell death occurs shortly thereafter, resulting in a decrease in turbidity measured at OD_{600}. During exponential growth the cells are buoyant and the culture is uniformly turbid. During plateau, the dead cells drop to the bottom and can be observed as a sediment, stirred up when the tube or flask is flicked. At the end, the cells are all on the bottom of the tube or flask and the culture media clears. Cell death occurs when nutrients become depleted and waste products build up in the media. Both protein and DNA are degraded after cells die, therefore it is important to catch the culture in the exponential growth phase.

In order to get the most out of gene expression and optimize yield, the length of time from induction to maximal protein production in the host *E. coli* cells must be determined. The induction mechanism is described in Experiment 14. The induction experiment will be prepared in advance by inoculating a small culture with cells from a single colony of *E. coli* BL21(DE3) cells carrying pET28LTB. When the culture has grown to exponential growth phase ($OD_{600} = 0.6$–0.8), the cells can be induced to begin protein expression by the addition of IPTG as described in Experiment 14. Aliquots are removed at specific times after IPTG addition. The aliquots can be frozen, then electrophoresed on sodium dodecyl-sulfate (SDS)–PAGE (Experiment 16) to determine the amount of new protein produced, and to estimate the time required for maximal new protein production. For some genes very good expression can be obtained within 2–5 h after induction. Expressing toxic or even slightly toxic gene products is a great problem with this system, however. Any background expression that impacts on the fitness of the host bacterium encourages natural selection of cells that produce lower levels of gene product. There may be highly variable expression when clones are selected for large-scale production of target protein (Pan and Malcom, 2000).

The pTZLT18R template plasmid isolated in Experiment 2 from which the LTB gene was cloned in Experiment 3 contains the entire operon encompassing the gene and all its control elements. It was derived from a plasmid isolated from the original pathogenic *E. coli* and cloned into *E. coli* DH5α (Cieplak *et al.*, 1995; Grant, Messer and Cieplak, 1994). The construct pTZLT18R and the pTZΔSA2B cloned later express LTB constitutively (continuously) and expression is therefore independent of IPTG induction in *E. coli* DH5α or BL21(DE3) cells (W. Cieplak, personal communication). The 13 preceding experiments have moved the LTB gene into a sophisticated plasmid system that allows much more flexibility for effectively manipulating both cloning and gene expression. If the previous experiments had not been successful, or if a control protein were desired, LT or LTB could be produced directly from the original constructs.

This induction experiment is therefore irrelevant if PTZLT18R is used. An SDS–PAGE gel of timed aliquots would show identical amounts of protein expression for each time because expression began before the time $= 0$ aliquot was taken.

Materials and Method

Materials

♦ Spectrophotometer set at 600 nm

♦ Shaker bath set at 37°C

♦ P-20 and P-200 Pipetmen, sterile white or yellow tips

♦ Plates with transformed BL21(DE3) carrying pET28LTB colonies (Experiment 14)

♦   15 ml culture tubes containing 3 ml LB-Kan30 broth

♦   80% (v/v) glycerol, autoclaved

♦   100 mM IPTG (isopropyl-β-D-thiogalactopyranoside) in sterile distilled water

♦   1.5 ml microcentrifuge tubes (one sterile, others need not be)

♦   SDS–PAGE sample loading buffer

Method

Done in advance of experiment: cell culture

(1) Using a sterile white tip, pick a colony from the transformant plate. Inoculate 3.5 ml LB-Kan30 broth. To inoculate: open the cap, flame the mouth briefly, drop the tip into the broth, flame again and close loosely.

(2) Grow the cells at 37°C, with shaking, until the $OD_{600} = 0.6$–0.8. Once the desired turbidity is reached, the culture may be stored at 4°C briefly, or overnight. This level of turbidity may take longer (up to 24 h) than the 3–6 h usually required due to the effect of pET28LTB on cell growth. Sufficient turbidity is required to provide enough cells for the subsequent SDS–PAGE. Clearing of the culture (decrease in turbidity) should be avoided as it indicates cell death. The cultures should not be overgrown at 37°C. This may deplete the media of antibiotic and allow re-growth of non-transformed cells. It is also possible that the *lac*UV5 promoter required for expression of T7 RNA polymerase may become unrepressed, allowing protein production before induction. In that case the SDS–PAGE gel of timed aliquots would show identical amounts of protein expression for each time samples were taken if expression began before the first aliquot was taken.

(3) Remove 0.5 ml of culture and place it in a sterile 1.5 ml microcentrifuge tube. Add 75 μl 80% glycerol. Label and store frozen for use in Experiment 18.

Time course for protein expression

(1) (a) If the cells have been stored for longer than a few hours, centrifuge them to sediment the cells and resuspend in 2 ml fresh media. Remove a 500 μl aliquot to a sterile microcentrifuge tube. Label it with '$t = 0$', the experiment number and the student ID.

Or

(b) If the cells have not been stored for more than an hour or two, remove a 500 µl aliquot to a microcentrifuge tube. Label it as described in step (1a).

(2) Collect the cells from the 500 µl aliquot by centrifugation (10 min at 10 000 *g*) and remove the supernatant. Discard the supernatant.

(3) Re-suspend the cell pellet in 50 µl 1× SDS–PAGE sample loading buffer and freeze at −20°C.

(4) Add IPTG to the remaining culture from a 100 mM IPTG stock for a final concentration of 1 mM (*x* µl×100 mM = 1500 µl×1 mM) for the T7 *lac* promoter in pET 28b and the *lac* promoter controlling T7 Polymerase production in the BL21(DE3) host.

(5) Place the tubes of induced cultures in a beaker or test tube holder to incubate in the shaker bath at 37°C.

(6) After 30 min, remove a 500 µl aliquot to a microcentrifuge tube. Label it with name, the date, and *t* = 30 min.

(7) Immediately repeat steps (2) and (3) on the aliquot removed.

(8) Sixty minutes after beginning the incubation, remove a third 500 µl aliquot and immediately treat as in steps (2) and (3).

(9) After 120 min from IPTG induction, remove a fourth 500 µl aliquot and immediately treat as in steps (2) and (3).

(10) Next day, remove another 500 µl aliquot and treat as in steps (2) and (3). *Be sure to note the time for this aliquot.* This last aliquot should give an indication of whether more or less time must be allowed for maximum protein production.

Preparation for experiment

(1) Done in advance of experiment: cell culture (see p 121).

(2) 100 mM IPTG stock (FW 238.3) Isopropyl-β-D-thiogalactopyranoside 0.238 g in 10 ml sterile distilled water.

(3) 1× SDS–PAGE sample loading buffer (Sambrook and Russell, 2000): 50 mM Tris–HCl, pH 6.8 (see Experiment 2 for instructions on making Tris buffers); 100 mM

dithiothreitol (DTT) added just before use*; 2% (w/v) electrophoresis grade SDS; 0.1% bromophenol blue; and 10% (v/v) glycerol.

Use 1 ml 500 mM Tris-HCl pH 6.8; 1 ml 1 M DTT 0.2 g SDS; 0.01 g bromphenol blue; 1 ml glycerol. Make up to 10 ml with distilled water. Aliquot 0.5 ml portions and store frozen.

References

Cieplak W. Jr, Mead, D. J., Messer, R. J. and Grant, C. C. R. (1995). Site-directed mutagenic alteration of potential active-site residues of the A subunit of *Escherichia coli* heat-labile enterotoxin. *J. Biol. Chem.* **270**, 30545–30550.

Grant, C. C. R., Messer, R. J. and Cieplak, W. Jr. (1994). Role of trypsin-like cleavage at arginine 192 in the enzymatic and cytotonic activities of *Escherichia coli* heat-labile enterotoxin. *Infect. Immun.* **1994**, 4270–4278.

Sambrook, J. and Russell, D.W. (2000). *Molecular Cloning: a Laboratory Manual*. Cold Spring Harbor, NY: Cold Spring Harbor Laboratory Press.

---

*Store 1× SDS–PAGE LB without DTT at room temperature. Add DTT from a 1 M stock just before the buffer is to be used or add 0.154 g directly to 10 ml prepared loading buffer. Several microliters of 2-mercaptoethanol may be added directly to the prepared 2× buffer in place of the DTT.

EXPERIMENT 16

(a) SDS–PAGE of Induction Time Course
(b) Transfer of Protein to Membrane for Western Blot

With this experiment we shift from molecular biology and DNA to the biochemistry and characterization of proteins.

Introduction

For cells induced at around $OD_{600} = 0.6$ and harvested 3 h later, Novagen claims a typical yield of around 0.5–1 mg total protein per milliliter of culture. Uninduced cells have about one-third this amount of total protein. The induced cells should therefore contain a large amount of target protein in addition to the normal protein components. About 10 µg of total protein are needed for a polyacrylamide gel lane on a mini-gel apparatus with 10-well comb, so one can estimate that 10–20 µl of induced cells and 50 µl uninduced cells should give proper protein band intensities after Coomassie Blue staining. The cells were collected, re-suspended in sample buffer and frozen in Experiment 15. The samples will be thawed and run on an SDS–PAGE to visualize the increase in target protein production. The same gel will be used to specifically verify the presence and band position of LTB.

LTB is a pentameric protein, the monomeric subunits being held together by strong electrostatic and some hydrophobic interactions. When heated in the presence of SDS, the pentamer dissociates into identical molecular weight monomers, 11 800 Da each. However, if the sample is not heated, the pentamer will remain together and travel as a

59 000 Da protein (Gill *et al.*, 1981). If the heated sample presents a single 11 800 Da monomer band and the primary band in the unheated sample migrates as a 59 000 Da protein one can reasonably expect that LTB has been successfully expressed. Nevertheless, the presence of LTB will be confirmed by an immunochemical method in Experiment 17.

SDS–PAGE

SDS–PAGE is the abbreviation for polyacrylamide gel electrophoresis with a sodium dodecyl sulfate-containing buffer. In this technique, proteins are separated in a poly-acrylamide gel medium by the application of an electrical field. See Experiment 12 for polyacrylamide structural information. *Always wear gloves when handling polyacryla-mide gels. The polymerized material is not toxic but unpolymerized neurotoxic acrylamide monomer may be entrapped in the gel.*

A non-denaturing PAGE gel separates proteins as a function of both charge and size using simple buffers and applied voltage. The more highly charged a molecule is, the faster it will be pulled through the gel by a constant voltage applied. However, the larger the molecular size (effective cross-section), the larger the frictional coefficient, so larger molecules will be retarded in the gel matrix based on molecular size and shape. These conflicting variables mean that determining the size or molecular weight of a protein based on mobility in a non-denaturing PAGE gel is problematic.

The problem is addressed by using a denaturing medium. In a denaturing gel, the buffer contains SDS, an anionic (negatively charged) amphipathic detergent. The dodecyl portion, $CH_3(CH_2)_{11}$, is hydrophobic. The sulfate, $-SO_4^{2-}$, carries two Na^+ counter-ions, making that portion of the molecule hydrophilic. When SDS binds a soluble protein, the hydrophobic portion of the SDS inserts into the hydrophobic core of the protein, causing the protein to denature by unfolding it. The Na^+ neutralizes electro-static interactions. The negatively charged sulfate is exposed and the protein is effectively covered with negative charges. SDS binds proteins in a constant mass ratio ($\sim$1.4 g SDS/g protein) and, therefore, in the presence of SDS, soluble proteins are denatured to a rod shape and possess roughly the same charge/mass ratio. Migration in an electric field depends on size, shape and charge. Since the denatured proteins have roughly the same shape and the charge is reduced to a uniform negative charge, they will migrate toward the anode (positive electrode) at a rate that is dependent only on the sieving action of the polyacrylamide matrix. Mobility of proteins subjected to SDS–PAGE was found empirically to be inversely proportional to molecular mass (Weber and Osborne, 1969). It should be noted that membrane proteins have a hydrophobic outer surface and do not bind SDS in the same way as aqueous-soluble proteins. Although soluble protein mobility is generally predictable, membrane proteins may have an unpredictable mobility in an SDS–PAGE gel. Since SDS denatures proteins by breaking the electrostatic interactions that maintain their conformations, it also dissociates non-covalently linked subunits into monomers, especially if the protein is heated in the presence of SDS.

Phosphate can interact with some proteins so the technique was modified and refined shortly after Weber and Osborne's paper by Laemmli (1970), who used a two-tiered gel system (discontinuous SDS–PAGE) and Tris–HCl/Tris–glycine buffers. Samples are generally further denatured by heating and the covalent disulfide bonds are reduced to thiols with reducing agents such as 2-mercaptoethanol (2-ME) or DTT (Clelland's reagent). These reagents reduce both intra- and intermolecular bonds (Cys–S–S–Cys to Cys–SH+Cys–SH) to allow complete denaturation. Dithiothreitol forms an insoluble product when it reduces disulfide bonds, so very little of it is required (Le Chatelier' Principle – a product removed from the reaction mixture drives the reaction to completion). The β-mercaptoethanol, on the other hand, has an equilibrium constant of 1, meaning the reaction goes forward and reverses at the same rate. A large excess is therefore required to drive the reaction to the right.

Acrylamide gels may be cast (polymerized) with a constant percentage acrylamide monomer and a set percentage cross-link, or may be cast in a gradient of percentages. Like agarose, acrylamide sieves molecules based on size, the mechanism being frictional drag through the gel matrix. The following percentages are recommended for separation: 3–5% acrylamide >100 000 Da; 5–12% acrylamide 20 000–150 000 Da; 10–15% acrylamide 10 000–80 000 Da; and 15% for molecules smaller than 15 000 Da.

Commercially obtained precast gels are recommended for this experiment to avoid handling of the potentially dangerous acrylamide monomer and for the sake of consistency. The gels may be simply 12% acrylamide, with a 4% stacking portion just beneath the well, or they can be 4–20% linear gradient gels, which provide excellent protein band resolution. In either case the gel buffer is Tris–HCl, and the running buffer is Tris–glycine/SDS.

Protein molecular weight markers are mixtures of several proteins found to migrate with precise mobility in the SDS–PAGE system. The composition of the mixture varies depending on the range of molecular weights used.

After electrophoresis, the separated protein bands are visualized by staining, usually with Coomassie Brilliant Blue (μg sensitivity) or a silver stain (ng sensitivity). The stain is applied in a solvent that serves as a fixative (precipitant) to immobilize the protein bands in the gel. Coomassie Brilliant Blue (an Australian name) is believed to interact electrostatically with lysine residues in a protein. It does not interact effectively with polyacrylamide so the gel is de-stained using dilute acetic acid, leaving the background clear and the blue dye associated with the protein bands. Sambrook and Russell (2000, p. A.46) provide an interesting note on the history of the dye.

Western blot

A Southern blot is performed with electrophoresed DNA transferred to a nitrocellulose membrane. The DNA bands are visualized on the nitrocellulose membrane using labeled hybridization probes. This technique was first described by Professor E. W. Southern, Department of Zoology, University of Edinburgh, Scotland (Southern, 1975). Later, the same transfer/hybridization technique, but using RNA, was given the nickname 'Northern blot' to distinguish it from Southern's DNA blot. Still later, the transfer of electrophoresed protein onto a nitrocellulose membrane with subsequent immunochemical detection was called a 'Western blot'. After electrophoresis, the proteins are transferred to a nitrocellulose membrane by capillary action and adsorption, then the bands can be visualized using anti-LTB antibody and an enzyme-labeled antibody detection system.

Another method to determine whether target protein is being expressed involves plating on a special agar formulation to indicate whether or not the colonies are expressing genes associated with the T7 RNA polymerase promoter. This procedure is an excellent way to quickly determine whether expression has taken place, but is non-specific with respect to the proteins expressed. The Western blot method provides specificity and an estimate of relative amount, although that evaluation depends on the efficiency of protein transfer to nitrocellulose.

Materials and Methods

Materials

Pre-blot electrophoresis

♦ Vertical polyacrylamide minigel electrophoresis apparatus* (gel size 8.6×6.8 cm $\times 1$ mm) with 50 μl wells

♦ Power supply

♦ P-200 Pipetman, gel-loading tips

♦ Precast 12% polyacrylamide gels or precast 4–20% acrylamide Tris—HCl gels

♦ $1 \times$ Tris—glycine/SDS running buffer (5 mM Tris/200 mM glycine/0.02% SDS, pH ~8.3; diluted from $5 \times$ buffer)

♦ Unstained or pre-stained broad-range molecular weight markers, 200 000–7000

---

*Any vertical PAGE apparatus may be used and assembled according to manufacturer's instructions.

◆ Coomassie Brilliant Blue stain (0.25% in 50% methanol/10% acetic acid solution) or a zinc stain–destain kit (BioRad catalog number 161-0440)

◆ 15% glacial acetic acid or 30% methanol/10% acetic acid for destain

Transfer blot

◆ Electrophoretic transfer cell (any other transfer apparatus may be used according to manufacturer's instructions)

◆ Magnetic stirrer

◆ SDS–PAGE gel

◆ Transfer buffer (25 mM Tris/192 mM glycine/20% v/v methanol, pH 8.3)

◆ Nitrocellulose membrane (BioRad catalog no. 162-0146)

◆ Filter paper cut to gel size

Method

Electrophoresis (SDS–PAGE)

The directions given below are for the BioRad Ready Gel Electrophoresis Cell (catalog no. 165-3125 or 165-3126).*

Assembly

(1) Remove the electrode assembly form the clamping frame. Rotate the cams outward to release the electrode assembly.

(2) Prepare a precast Ready Gel by removing the tab from the bottom of the cassette. Repeat for a second gel. If only one gel is run, place the buffer dam in the empty slot.

(3) Place the cassettes into the slots at the bottom of each side of the electrode assembly. Be sure that the short glass plate of the Ready Gel cassette faces *inward* toward the notches on the green U-shaped gaskets.

(4) Press the Ready Gel cassette up against the gaskets. These fit together to create the inner chamber.

(5) Transfer the electrode assembly and gels into the clamping frame.

---

*Reproduced by permission of BioRad.

(6) Press down on the electrode assembly with forefingers of each hand while simultaneously closing the two cam levers of the clamping frame with the thumbs of each hand.

(7) Lower the electrode assembly and clamping frame into the minitank.

(8) Fill the inner chamber with 125 ml of running buffer, so that the buffer reaches a level between the tops of the short and long plates of the Ready Gel. Do not overfill.

(9) Add 200 ml of running buffer to the minitank.

Sample preparation

(1) Divide each sample into two tubes. Heat one tube to 90°C for 4 min in a beaker filled with water. Mark it 'H' and the time after induction. Leave the other half of the sample at room temperature. Mark it 'RT' and the time after induction.

(2) Load 40 µl, $T = 0$ (uninduced) sample/well. Load 20–30 µl of each subsequent time sample/well, precisely noting the contents of each well. Load the heated samples next to one another for comparison as time after induction increases.

(3) Repeat with the unheated samples. Load one lane of each gel with a protein molecular weight marker.

Run the gel

(1) Align the electrode plugs and jacks and place the lid on top of the minitank. Match colors of the plugs on the lid with the jacks on the electrode assembly.

(2) Attach the electrical leads to a suitable power supply (150 V minimum) with the proper polarity (red—red, black—black).

(3) Electrophorese for 35–45 min, at 150–200 V constant (60 mA current) for one gel or two until tracking dye is 1–2 cm from bottom.

   Note: while the gel is running, prepare transfer filters and membranes; make sure the buffer is ice-cold and Bio-Ice is frozen (see below).

(4) When the dye is near the bottom of the gel, turn off current, unplug the power supply then remove the gel. *Use gloves so stainable fingerprints are not left on the gel.*

Stain one gel

At least one gel containing a time series should be stained with Coomassie Blue or zinc stain to see whether expression has, in fact, occurred. This provides a picture of all the bacterial proteins being expressed and an estimate of the proportion of the total represented by the target protein (LTB).

(1) Transfer the gel to the stain bath and leave overnight. Gentle rocking during staining minimizes precipitation of undissolved dye onto the gel.

(2) Destain next day by placing the gel into destain solution and gently rocking the container. Destaining may be speeded up by including a few grains of anion exchange resin or a sponge in the destain solution and changing the solution two or three times.

(3) When destained, transfer the gel to 20% glycerol in water to prevent shrinkage and cracking as it dries. Soak for 1 h or more. The gel may be transferred, in glycerol solution, to a plastic bag for storage at 4°C, or it can be dried in air at room temperature clamped between two glass plates.

Note: storage in destain solution will eventually cause the bands themselves to destain. For zinc stain follow manufacturer's instructions.

Transfer protein to nitrocellulose

Instructions are for a BioRad Mini Trans-Blot Electrophoretic Transfer cell *

Prepare buffer in advance and refrigerate. The buffer temperature must be 4°C at the start of transfer. Freeze the BioIce unit in advance.

(1) Following electrophoresis, transfer the remaining gels to a container filled with transfer buffer and gently wash the gel using a side-to-side motion. This equilibrates the gel with transfer buffer and allows the gel to retain its final size, as well as removing salts and SDS that would have altered conductivity and result in heat generation.

(2) If it is not already pre-cut, trim the membrane to the dimensions of the gel. Use gloves to handle the membrane. Label the membrane with a soft pencil to identify the gel and the orientation of the membrane.

(3) Wet the membrane by slowly sliding it at a 45° angle into the transfer buffer and allowing it to soak for 15–30 min. The membrane must be completely wet and all air bubbles squeezed out. *Use forceps and wear gloves to avoid transferring proteins from fingers to the membrane.*

(4) Completely saturate the pre-cut filter paper and filter pads by soaking them in transfer buffer. Avoid entrapping air bubbles. Use thick filter paper.

*Reproduced by permission of BioRad.

(5) Thoroughly rinse the buffer chamber with distilled water to remove SDS from previous use. Install the Mini Trans-Blot electrode in the buffer chamber. Fill the buffer tank half full with transfer buffer (~400 ml) and install a 1 in stir bar at the bottom of the unit.

Assembly

Wear gloves to avoid contaminating the membranes.

(1) Open the gel holder cassette by sliding and lifting the latch. The clear panel is the anode (+) side and the gray panel is the cathode (−) side. For the electrode module, the cathode is the black electrode panel located in the center of the buffer tank. Always insert the gel cassette so that the gray plastic faces the black plastic of the cathode electrode. The anode and cathode are black (cathode) and red (anode) sides of the electrode module.

(2) Place the opened gel holder in a shallow glass dish so that the gray panel is flat on the bottom of the dish. The clear panel should be at an angle against the wall of the dish.

(3) Place a pre-soaked fiber pad on the gray panel of the cassette. Be sure to center all components, otherwise the transfer pattern will be distorted.

(4) Place a piece of saturated filter paper on top of the fiber pad. Saturate the surface of the filter paper with 2–3 ml of transfer buffer.

(5) Place the equilibrated gel on top of the paper.

(6) Align the gel in the center of the cassette. The gel must be within the pattern of circles in the holder. Make sure no air bubbles are trapped between the gel and filter paper.

(7) Flood the surface of the gel with transfer buffer and lower the pre-wetted blotting media on top of the gel. Do this by holding the membrane at opposite ends and bending it so that the center portion will contact the gel first. Gradually lower the ends. Next, roll a glass pipet or test tube over the top of the membrane (like a rolling pin) to exclude air bubbles from the area between the gel and membrane. Apply pressure until a nearly adhesive contact is made between membrane and gel.

Note: failure to obtain good contact will result in swirled or missing transfer patterns and overall high background. If good contact is made, but air bubbles remain, the bubble patterns will transfer to the membrane.

(8) Flood the surface of the membrane with transfer buffer. Complete the sandwich by placing a piece of saturated filter paper on top of the membrane and placing a saturated filter pad on top of the filter paper.

(9) Close the cassette. Hold it firmly so that the sandwich will not move, and secure the latch.

(10) When the gels to be transferred are in place, set the buffer tank on top of a magnetic stirrer. Fill the tank with transfer buffer to just above the level of the top row of circles on the gel holder cassette. Do not overfill.

(11) Turn on the magnetic stirrer and put the lid in place. Be sure the electrode wires on the lid are attached to the appropriate pins on the electrode module (black wire to cathode, red wire to anode).

(12) Plug the unit into the power supply. Transfer is from cathode to anode.

(13) Turn on the power supply to initiate transfer.

Transfer conditions for Tris – HCl / Tris – glycine SDS gels

◆ Blotting buffer is transfer buffer: 25 mM Tris/192 mM glycine/20% v/v methanol

◆ Media: nitrocellulose

◆ Power conditions: with pre-frozen Bio-Ice unit in place, 650 ml transfer buffer at 4°C; rapid transfer (1 h) starts with 100 V, 250 mA and ends with 100 V, 350 mA; overnight transfer starts with 30 V, 40 mA and ends with 30 V, 90 mA.

An overnight transfer is convenient if the laboratory session begins late in the day. Apparatus is turned off the next morning, and membranes are removed and stored at 4°C for later visualization.

Preparation for experiment

(1) Running buffer: 5× Tris–glycine SDS buffer stock (25 mM Tris/1 M glycine/0.1% SDS, pH ~8.5) – 1.5 g Tris base; 37.5 g glycine; 0.5 g SDS. Make up to 500 ml with distilled water. Check pH and adjust if necessary. 1× Tris–glycine/SDS running buffer – 5 mM Tris/200 mM glycine/0.02% SDS, pH ~8.3; diluted 1:5 from 5× buffer.

(2) Stain solution (methanol:acetic acid), 50% methanol, 10% acetic acid: 500 ml methanol; 400 ml water; 100 ml glacial acetic acid. Dissolve 0.25 g Coomassie Brilliant Blue R-250 in 100 ml methanol:acetic acid solution. Filter the solution through Whatman no. 1 filter paper to remove undissolved material. The solution

can be recycled after use on a non-denaturing gel. If an SDS gel is to be stained, the gel can be soaked for a few minutes in methanol:acetic acid solution to remove the SDS and immobilize the protein prior to immersion in Coomassie Blue solution, which can then be recycled.

(3) Destain (30% methanol, 10% acetic acid): 600 ml distilled H_2O; 300 ml methanol; 100 ml acetic acid.

(4) Prepare the Bio-ICE cooling unit in advance: fill with deionized distilled water and freeze. Install the frozen cooling unit in the buffer chamber next to the electrode a few minutes before beginning transfer.

(5) Blotting buffer: transfer buffer. Buffer for SDS—PAGE protein transfers: 25 mM Tris, 192 mM glycine, 20% v/v methanol; 3.03 g Tris, 14.4 g glycine, 200 ml methanol, distilled deionized water to 1 l. The pH will range between 8.1 and 8.4 depending on the quality of the reagents. Methanol must be analytical grade as metal contaminants will plate on the electrodes. The buffer must not be adjusted by addition of acid or base. Improperly made or adjusted buffer will cause excess heat generation.

Note: Zinc stain is easier and more sensitive but is more expensive to use.

References

Gill, D. M., Clements, J. D., Robertson, D. C. and Finkelstein, R. A. (1981). Subunit number and arrangement in *Escherichia coli* heat-labile enterotoxin. *Infect. Immun.* **33**, 677–682.

Laemmli, U. K. (1970). Cleavage of structural proteins during the assembly of the head of bacteriophage T4. *Nature* **227**, 680–685.

Sambrook, J. and Russell, D.W. (2000). *Molecular Cloning: a Laboratory Manual*. Cold Spring Harbor, NY: Cold Spring Harbor Laboratory Press.

Southern, E. (1975). The detection of specific sequences among DNA fragments separated by gel electrophoresis. *J. Mol. Biol.* **98**, 503–516.

Weber, K. and Osborne, M. (1969). The reliability of molecular weight determinations by dodecyl sulfate-polyacrylamide gel electrophoresis. *J. Biol. Chem.* **244**, 4406–4412.

EXPERIMENT *17*

Visualization of Western Blot

Introduction

Enzyme-linked immunosorbent assay (ELISA) is one of the most sensitive and specific assays for identifying proteins. It originated with the development of radioimmunoassay (RIA) techniques, and represents a non-radioactive modification of that method. There are many variations of ELISA that can be designed to explore many kinds of research questions, so it has proved a most valuable and extensively employed procedure. It is the basis for pregnancy kits, HIV detection and quite a few clinical assays. In this experiment it is performed as a solid-phase immunological assay, meaning that the proteins to be assayed are immobilized on a solid surface, in this case nitrocellulose. The same solid-phase ELISA technique will be used in a 96-well plate format in Experiment 23 to assay for ligands that are recognized by LTB. An antibody (primary antibody) specifically recognizes and binds to the target protein (antigen), LTB. The antibody was elicited by the immune response of an animal challenged with LT or LTB antigen. For the Western blot, the protein has been transferred to a membrane where it remains immobilized. The protein and the membrane are treated with a non-specific blocking agent such as milk proteins of bovine serum albumin (BSA) to block any non-specific sites and background surfaces that might be bound with low affinity by primary antibody, producing a background signal. After blocking, only antigenic sites on the protein that bind with high affinity to primary antibody will be detected quantitatively.

The target protein LTB, being a portion of the LT holotoxin, is recognized and thus identified by the primary antibody, a rabbit-derived anti-LT polyclonal antibody. The LTB protein is also recognized, although to a slightly lesser extent, by polyclonal antibodies against CT (cholera toxin) and CTB (cholera toxin B subunit, also called 'choleragenoid'). This effect is called cross-reactivity. To determine how much primary antibody has bound to the LTB antigen a secondary antibody raised against the constant region of the primary antibody is employed. The constant region of an antibody contains amino acid sequences that are characteristic of the host animal in which the antibody has been elicited. Rabbits were immunized with LT and the antibody they produced, called rabbit anti-LT immunoglobulin G (IgG) therefore contains constant region amino acid sequences characteristic of rabbits. The secondary antibody for this system was raised by immunizing goats with rabbit-derived IgG, and is therefore called goat anti-rabbit IgG. Anti-CTB is elicited in goats, therefore the secondary antibody used for anti-CTB is rabbit or mouse anti-goat IgG. The secondary antibody is purified and then conjugated (covalently linked) to an enzyme, horseradish peroxidase (HRP), which catalyzes an electron transfer reaction between peroxides and an electron donor substrate. HRP catalyzes the reduction of peroxides to water. Reaction with peroxide oxidizes the HRP enzyme, which now requires an electron donor to reduce it back to its native state. Since the reaction is a two-electron transfer, two donor molecules are required to return the enzyme to its original oxidation state. The donor substrate used for this procedure is a chromophore. In the reduced state the chromophore is colorless. After donating an electron to oxidized HRP, the oxidized chromophore becomes a colored solution or precipitate.

In this way LTB antigen is detected, identified and relatively quantitated by specific binding of primary antibody (rabbit anti-LT IgG) to LTB then measuring the reaction product obtained when a secondary antibody coupled to horseradish peroxidase (goat anti-rabbit IgG-HRP) binds the primary antibody then reacts with a peroxide substrate and an electron transfer chromophore. A precipitating chromophore such as 4-chloro-1-naphthol reagent containing peroxide substrate is used to localize the site of the bound protein band on the nitrocellulose. Other precipitating chromophores are equally effective, including diaminobenzidine (DAB) and 3,3′4,4′-tetramethylbenzidine (TMB); chromophores that phosphoresce, luciferin, for example, are also suitable. Figure 17.1 shows the method graphically.

One blot will be stained to obtain an estimate of the relative amounts of protein that have been transferred to the nitrocellulose and as a comparison/control for the immunological assay.

Materials and Methods

Materials

- Rocker table, if available

- Small container, such as tip box cover or Petri dish.

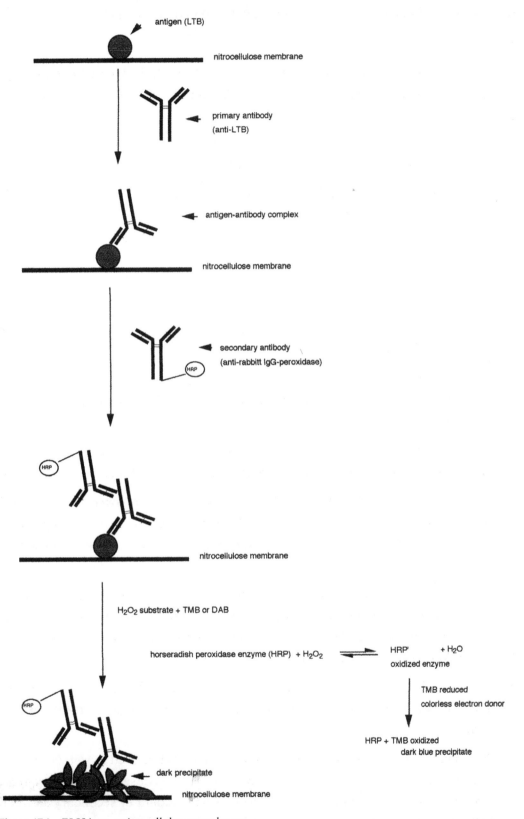

Figure 17.1 ELISA on a nitrocellulose membrane

- Tris-buffered saline (TBS; 20 mM Tris–HCl, 500 mM NaCl, pH 7.5)

- TBS/1% BSA

- Primary antibody (1:1000 rabbit anti-LT IgG in TBS/1% BSA or 1:1000 anti-CTB (goat anti-choleragenoid, List Laboratories, Campbell, CA, USA catalog no. 703) in TBS/1% BSA)

- Secondary antibody (1:1000 Goat anti-rabbit IgG-HRP conjugate in TBS/1% BSA, or 1:1000 rabbit anit-goat IgG-HRP if anti-CTB is used)

- One-step TMB-blotting or other one-step precipitating product (see Preparation for Experiment section for alternatives)

- Western blot from Experiment 16

Method

(1) Place the membrane in a small container. Cover with TBS to re-hydrate it. Soak for 5 min.

(2) Pour out TBS and replace it with TBS/1% BSA to at least 1 in above the membrane to block adsorption of primary antibody to the walls of the container. Soak for 30 min.

(3) Pour out TBS/1% BSA and add 1:1000 primary antibody to cover. Cover the dish with plastic film to minimize evaporation. Incubate for 45 min at room temperature, rocking the container frequently (see Experiment 23 for discussion of incubation times).

(4) Pour out primary antibody. Wash three times: add buffer to cover, rock side-to-side gently three or four times, pour out buffer and replace with another volume of buffer. Repeat rocking. Pour out second wash and repeat with a third volume of buffer. Work quickly as the membrane must stay moist.

(5) Pour out a third volume of wash buffer. Add 1:1000 anti-rabbit IgG-HRP (secondary antibody) to cover. Cover the dish with plastic wrap. Incubate for 45 min with rocking.

(6) Pour out secondary antibody and wash three times, as described in step (4). For the last wash, fill the container to the top to be sure all traces of secondary antibody are removed. Effective washing is crucial. Even a small amount of secondary antibody will react with the chromophore and produce background color.

(7) Pour out last wash solution and immerse the membrane in pre-mixed chromo-phore−substrate solution. Protein concentrations of at least 100 ng will become visible immediately as a blue band with TMB.

(8) Stop color development by immersing the membrane in distilled water for 10 min. Change the water after the first 5 min.

(9) Record observations.

Preparation for experiment

(1) Tris-buffered saline (TBS): 20 mM Tris−HCl/500 mM NaCl, pH 7.5 (see Experiment 2 for Tris buffer preparation). Buffer should be 0.2 μm filtered.

(2) TBS/1% BSA: 20 ml 10% BSA in water made up to 200 ml with TBS. TBS is recommended because it can be used with either HRP or alkaline phosphatase (AP) chromophore systems. The AP is inhibited by phosphate buffer. The BSA is generally made as a 10% stock (5 g BSA+45 ml water), 0.2 μm filtered and kept at 4°C to reduce bacterial growth. It is diluted to 1% for use and the 1% BSA is discarded after use as it becomes contaminated very easily. To dilute, add 5 ml of 0.2 μm filtered 10% BSA in water to a 50 ml sterile centrifuge tube and make up to 50 ml with 0.2 μm filtered TBS.

(3) 1:1000 rabbit anti-LT IgG and 1:1000 goat anti-rabbit IgG-HRP conjugate.

Note: if anti-LT is not available, anti-CTB can be obtained commercially from List Labs Inc., Campbell, CA, USA, catalog number 703, or from Sigma-Aldrich (St Louis, MO, USA). Concentrated (1 mg/ml) re-suspended antibody should be distributed into 100 μl aliquots and stored frozen. If anti-CTB is used, the secondary antibody must be anti-goat IgG-HRP.
1:50 stock: add 100 μl of antibody to 5 ml TBS/1% BSA. The 1:50 stock can be stored at 4°C; 1:1000 working solution: just prior to use, add 500 μl of 1:50 stock to 10 ml TBS/1% BSA.

(4) Alternative chromophore formulations:

♦ DAB/NiCL$_2$ forms a dark brown precipitate and is quite sensitive. The benzidine is potentially carcinogenic, however.

♦ TMB forms a dark purple stain. It is more stable and less toxic than DAB. It can be used with all membrane kits and can be obtained as a stable 'one-step' formulation containing chromophore and a urea-peroxide, thus eliminating the requirement for mixing with 30% H_2O_2 substrate prior to use.

Both DAB and TMB with or without peroxide substrate can be obtained from Pierce Chemical (Rockford, IL, USA); Sigma-Aldrich (St Louis, MO, USA); Kirkegaard & Perry (Rockville, MD, USA) and several other suppliers. If substrate is to be added separately, the recipe is given or the substrate supplied with the chromophore.

♦ Luminol/H_2O_2/p-indophenol (HRP based): oxidized luminol gives off a blue light. The p-indophenol enhances light output. This and other chemiluminescent systems are available from the suppliers listed above. They require X-ray film for a permanent data record.

The HRP systems are recommended because they are more robust and reliable than alkaline phosphatase-based systems. Alkaline phosphatase is inhibited by phosphate buffers but can be used according to the manufacturer's protocol if TBS is used exclusively.

EXPERIMENT *18*

Large Scale Culture Preparation

Note: This is not, strictly speaking, a laboratory session. The manipulations must be done over a period of several days and the logistics depend on individual situations.

The large scale culture can be planned as a single large culture to be distributed to individual students for use in Experiment 19 as 5–10 ml aliquots after harvest and sonication. Approximately 100–250 ml media per team would be required. Alternatively, a series of 100 ml cultures can be grown, induced and harvested. Instructions are given in detail and should be adapted to fit particular requirements and time constraints.

Introduction

The time has come to prepare a large quantity of recombinant protein for biochemical studies. To accomplish this goal, several liters of *E. coli* BL21(λDE3) carrying the pET28LTB construct are grown up, cells are harvested and broken open to release their contents, including LTB protein. Large-scale growth is best initiated by a series of steps leading up to inoculation of 300–500 ml of culture media with approximately 50 ml of actively growing cells. Bacteria grow by fission, doubling their number. At 37°C, the doubling time may be as short as 20–30 min. This creates an exponential growth curve, beginning with a lag time due to induction of new proteins necessary for growth, the synthesis of additional cell components and to some extent the difficulty of measuring very small numbers of cells as they begin to double from a small inoculum. The organizational period is followed by rapid doubling, the exponential growth phase. Eventually, various factors such as reduction in nutrient supply and accumulation of waste products that may be toxic, cause the growth curve to plateau. In the plateau phase, just about as many cells die as are produced. Both proteins and DNA are degraded after death. Active growth in the exponential phase is the best time to obtain maximum protein with minimum dead cells in the culture.

Occasionally a protein is poorly expressed, even in a strong expression system. This may be due to toxicity of the protein for the host cell. In that case, cell growth continues for some time after induction of the protein, but then the cells lyse, evidenced by a decrease in the cell density measured at OD_{600}. In that case, cells must be harvested a few hours after induction to allow sufficient time for satisfactory quantities of recombinant protein to accumulate before the protein level becomes toxic to the host. Thus, time of harvest is crucial. Expression of LTB should not present toxicity problems since it is an *E. coli* protein without enzymatic activity. Overexpression of LT or LTA may have consequences for the T7 RNA polymerase system, as discussed in Experiment 15.

The final inoculum is grown up due to the effect of increased cAMP production on the host bacterial cell to an $OD_{600} = 0.4–0.6$ (cells in exponential phase, actively growing) then induced with IPTG. The previous tests during Experiments 15 and 16 will have determined the optimum time to stop growth after induction. To minimize the amount of material to be processed and facilitate cell breakage, cells are pelleted by centrifugation and the pellet is frozen at $-70°C$. After several hours, up to a few days, the pellet is thawed, breaking open many of the cells and allowing their contents to spill out. This is necessary because *E. coli*, like many Gram-negative bacteria, does not actively secrete protein products into the media. The proteins, either endogenous or introduced, are generally secreted into the periplasm of the bacterial cell, the space between the bacterial plasma membrane and the cell wall. The thawed pellet is resuspended by vigorous vortexing in a small amount of Tris buffer containing lysozyme prior to submitting the cells to sonication. Lysozyme is a protein found in human tears, and in hen (or other avian) egg white. The lysozyme protein pierces bacterial cell membranes, weakening them. Sonicating the cells further disrupts the membranes and causes cell contents to be released into the surrounding medium, leaving cell debris.

Sometimes a recombinant protein is expressed in an insoluble form, known as an inclusion body, an aggregate deposited in the bacterial cytoplasm. This is particularly common when the protein is expressed at high levels. It is the usual bacterial method for storing excess metabolic products. The protein found in an inclusion body is generally thought to be mis-folded, and may therefore have to be solubilized, then re-folded, after separation from soluble proteins.

In place of sonication, several reagent mixtures are commercially available for protein extraction. These contain a nonionic detergent that helps to lyse the cells. One can obtain a 'soluble' and an 'insoluble' fraction, which can be used to determine whether recombinant protein is soluble in the cytoplasm or insoluble in an inclusion body. The disadvantage of these extraction reagents is that they may interfere with an affinity column used for isolation of the target protein. Dialysis before isolation is necessary in that case.

Materials and Methods

Materials

Culture preparation

- Autoclave

- Glycerol stock of BL21(λDE3)/pET28LTB from Experiment 15 or transformant plates from Experiment 14 with sequenced colony marked

- 1 l Erlenmeyer (conical) flask

- 125 ml Erlenmeyer (conical) flask

- 15 ml sterile culture tubes

- Inoculation loop

- Sterile toothpick or pipet tip

- 100 mM IPTG in sterile distilled water

- Bactotryptone

- Yeast extract

- NaCl or premixed Millers L-B Broth

Harvesting

- Probe sonicator (instructions are given for a Branson Sonifier, model 450)

- Preparative refrigerated centrifuge, JA 14 rotor (if not available, a table-top centrifuge that holds 50 ml tubes can be used)

- Centrifuge bottles, preferably 250 ml for large volume culture

♦ Cultures grown to $OD_{600} = \sim 0.6$

♦ Lysozyme stock (20 mg/ml in H_2O) (Sigma-Aldrich catalog number 6876)

♦ TEAN buffer: 50 mM Tris–HCl/pH 7.4, 1 mM EDTA, 3 mM NaN_3, 200 mM NaCl

♦ Sterile 50 ml centrifuge tubes

♦ Ice and ice bucket

Method

Media preparation

(1) Add 5 g Bactotryptone, 2.5 g yeast extract and 5 g NaCl or the requisite amount of premixed formulation to a 1 l flask, add 500 ml distilled water and swirl to disperse solids. Make one batch of 500 ml.

(2) Swirl the mixture and immediately dispense 50 ml media to one 125 ml conical flask.

(3) Cover flasks with aluminum foil. Add a piece of autoclave tape across the top. Autoclave 20–30 min using the liquids cycle.

(4) Remove flasks and place in a room temperature water bath. Cool to 60°C (cool enough to handle).

(5) Remove foil and quickly add 50 μl Kanamycin (30 mg/ml stock) to the 50 ml aliquot and 450 μl to the 450 ml of media for a final concentration of 30 μg/ml media. Re-cover flasks with the sterile foil. Swirl to mix. Avoid bubbles.

If using previously frozen glycerol stock from Experiment 15:

(6) Light the Bunsen burner and adjust the flame to blue, and about 2 in (4 cm) high.

(7) Remove the foil from the small culture flask. Flame the top of the flask.

(8) Add 0.3–0.5 ml of thawed glycerol stock from Experiment 15 [BL21(DE3)/pET28LTB] to the small culture flask. Swirl to mix.

(9) Flame the top again, re-cover the flask with the sterile foil and place it in a 37°C water bath shaker to incubate.

If using transformant plate [BL21(DE3)/pET28LTB] from Experiment 14 for initial inoculum: follow steps 7a–8a.

(7a) Flame the inoculation loop and pick a colony from the transformant plate. Open the 125 ml flask and swish the loop in the broth.

(8a) Flame the top of the flask, re-cover the flask with the sterile foil and place the flask in a 37°C water bath shaker to incubate.

Alternatively, wearing gloves, pick a colony with a sterile toothpick or sterile tip and drop it into the broth. This is more dangerous as hands may contaminate the pick.

(10) Incubate 3–6 h or overnight if necessary until $OD_{600} = \sim 0.6$. The flask may then be refrigerated for a few hours after reaching 0.6, if necessary.

(11) Open the 125 ml flask containing the fresh culture, and flame the top. If desired, remove 0.5 ml of culture to a sterile 1.5 ml microcentrifuge tube, add 75 μl 80% glycerol store at −70°C. Open the 1 l flask containing 450 ml LB-Kan30 and pour the remainder of the 50 ml culture directly into the fresh broth. Flame the lip and close the flask.

(12) Incubate in a 37°C water bath shaker until $\sim OD_{600} = \sim 0.4–0.6$.

(13) Add 4.5 ml of 100 mM IPTG stock to a final concentration of 1 mM. Swirl to mix.

(14) Incubate with shaking for the pre-determined optimal time at 37°C. It may be advantageous, instead, to incubate with shaking at 25°C overnight or, if necessary, 24–48 h. The lower temperature and longer time reduces inclusion body formation.

Note: Good results have been obtained from pTZLT18R inoculated from a colony or 0.5 ml glycerol stock into 300–500 ml LB-Carb50 (50 μg/ml Carbenicillin) broth and grown at 37°C with shaking to OD = 0.6, then induced with IPTG and allowed to grow at 37°C with shaking for up to 48 h. Several milligrams of LT protein were recovered. pTZLT18R carries the Amp resistance gene. Use Ampicillin or Carbenicillin selection.

Harvesting

(1) Centrifuge cultures for 5 min at 5500 *g* in a JA14 rotor at 4°C to sediment the cells, 10 min for larger volumes. Discard the supernatant and freeze the cell pellets for 6–24 h at −70°C.

(2) Thaw pellets. Resuspend them and pool in 25–30 ml TEAN buffer containing 50 μl of a 20 mg/ml lysozyme stock per original volume of 250–300 ml. Use proportionately less for 100 ml cultures. Place resuspended material in sterile 15 ml centrifuge tubes.

(3) Incubate resuspended pellets for 30 min at room temperature to allow lysozyme to begin cell lysis.

(4) Place the tubes on ice and sonicate twice for 2–5 min each time, depending on volume in a Branson Sonifier model 450 equipped with a large horn. Set the duty cycle to 50% and the power output at 7. The conditions are standard for *E. coli*.

(5) Following sonication, pellet the cell debris by centrifugation at 4°C for 15 min at 5500 *g* (16 000 rpm in a Sorvall refrigerated preparative centrifuge, in an SS34 rotor).

(6) Recover the supernatant. It may be stored at 4°C for several hours or frozen at −20°C. Discard the pelleted debris.

Preparation for experiment

(1) 100 mM IPTG: 0.238 g IPTG dissolved in 10 ml sterile distilled H_2O.

(2) TEAN buffer: 50 mM Tris–HCl/pH 7.4, 1 mM EDTA, 3 mM NaN_3 and 200 mM NaCl.
    6 g Tris base; 900 ml distilled water. Adjust the pH to 7.4 with concentrated HCl; add 1 ml 0.5 M EDTA stock (see Experiment 2); add 1 ml 3 M NaN_3 stock (1.95 g/10 ml distilled water) or 3 ml 1 M NaN_3 (0.65 g/10 ml); add 11.7 g NaCl. Make up to 1 l with distilled water.

(3) 20 mg lysozyme in 1 ml distilled water. Sigma-Aldrich catalog number 6876 is suitable.

EXPERIMENT *19*

Isolation of LTB by Affinity Chromatography

Introduction

Affinity chromatography is a type of liquid chromatography in which the molecule to be purified is specifically and reversibly adsorbed by a complementary binding substance, called a ligand, immobilized on an insoluble support (matrix). Purification from a complex mixture is greatly simplified because the affinity ligand can be chosen to bind preferentially or specifically to the target molecule, while the remaining components of the mixture are washed through the column. The procedure has a concentrating effect which enables large amounts of sample to be processed. Purification is often on the order of several-thousand-fold, removing very small amounts of biological material from large amounts of contaminating material, the recovery of active material is very high without the necessity for transient denaturation or harsh reagents. In addition, this one-step purification procedure eliminates time-consuming, complex, multiple-step purification.

The first application of affinity chromatography was in 1910, the selective adsorption of amylase protein onto insoluble starch. Finding or synthesizing complex matrices having covalently attached (immobilized) ligands prevented the technique from becoming generally available. The chemistry to covalently attach amino groups to a polysaccharide matrix activated by cyanogen bromide opened the way for general use. In addition, specific types of matrix supports were found that served as ligands for particular target molecules. Since LT has been shown to bind to the terminal galactose on ganglioside GM_1 (Merritt *et al.*, 1994; Tsuji *et al.*, 1985) an immobilized galactose column makes a great deal of sense for isolation and purification of the native toxin or B subunit, and Uesaka *et al.* (1994) has taken advantage of that fact. His protocol included purification from 6000 ml of culture, which was sonicated and cell debris discarded. The resulting supernatant was treated with 65% ammonium sulfate to precipitate the bulk of the proteins. The precipitate was resuspended in 21 ml buffer and run on 15 ml of immobilized galactose resin, resulting in isolation of about 13 mg pure LT eluted with 0.3 M galactose.

This experiment follows essentially the same method, although with smaller quantities of culture, sample and resin, aliquotted for each group. If Uesaka concentrated 6000 ml to 21 ml, which was run on a 15 ml column, the isolation should be feasible from 300 ml concentrated to 3 ml, and chromatographed on a 1 ml column.

The column eluent is evaluated by reading absorbance of the fractions at 280 nm. This is the wavelength absorbed by tyrosine and tryptophan in proteins. Phenylalanine also absorbs at approximately that wavelength, but its extinction coefficient (molar absorptivity) is too small to contribute to the total absorption. A protein with an unusual distribution of Tyr or Trp may give inaccurate estimates of total protein concentration. This is, in fact, the case with LTB and CTB, which contain a normal number of Tyr residues, but only a single Trp (Trp88) per monomer (total five Trp per molecule; Dallas and Falkow, 1980). A standard protein curve using BSA would indicate less LTB than is actually present.

Materials and Methods

Materials

- Spectrophotometer set at 280 nm

- 1 ml quartz cuvettes or tubes

- P-1000 Pipetman, sterile blue tips

- 15 ml centrifuge tubes

- 2 ml microcentrifuge tubes

- Immobilized galactose (Pierce Chemical, Rockford, IL, USA catalog no. 20372)

- Small disposable columns; may be plastic with frits or short Pasteur pipettes with glass wool insert

- TEAN buffer: 50 mM Tris–HCl/pH 7.4, containing 1 mM EDTA, 3 mM NaN$_3$ and 200 mM NaCl

- 1 M D(+)-galactose in TEAN buffer

- 0.3 M (+)-galactose in TEAN buffer

- 95% ethanol to rinse cuvette

Method

(1) Place a filter frit into a disposable column. Wet the frit with 5 ml TEAN buffer and allow the buffer to flow through, pulling with a slight vacuum if necessary. To pull through, attach a rubber bulb to the bottom of the column, expel the air then allow gentle suction to pull the liquid through.

(2) Suspend the immobilized galactose resin by gently inverting and swirling the bottle. Quickly remove 2 ml of suspension with a Pasteur pipette P-1000 Pipetman and place the suspension in the column. Allow the suspension to settle. Collect the buffer that drains away in a beaker or tube. Discard this buffer. Resin is supplied as a 50% suspension, so 2 ml should result in a 1 ml resin volume.

(3) Calculate your column volume by measuring the diameter of the column in centimeters. Use an extra frit for this purpose – it will be the same diameter as the column. Calculate the radius (r). Measure the height (h) of the resin column in centimeters. Calculate the column volume: $V = \pi r^2 h$. Units calculated in centimeters yield cm^3 = volume in ml.

(4) Wash the resin with 5 volumes TEAN buffer. That is, if the volume of resin is 1 ml, then fill the column with 5 ml TEAN buffer and allow the buffer to drain through. This removes any preservatives or other materials that may be in the resin suspension and equilibrates the column in TEAN buffer.

(5) Place the processed culture supernatant after sonification in the column. Up to 5 ml can be accommodated on 1 ml of resin. Collect the flow-through (eluent) in 3–5 ml aliquots, in 15 ml centrifuge tubes, numbered and labeled 'flow-through'. Check and record absorbances at 280 nm in the spectrophotometer. If they are over 3 record 'out of range'.

(6) Add additional TEAN buffer and collect 3–5 ml samples in 15 ml tubes until eluent runs clear. Continue washing with TEAN and collecting aliquots until absorbance is baseline (0.003, preferably less). You may require 12–15 ml total. Plot absorbance

(*y*-axis) vs fraction number or volume (*x* axis). See Uesaka *et al.* (1994) for example of how the plot should look. The tubes should be sequentially numbered.

(7) Add one to two more column volumes, collect this wash in a fresh tube and check the absorbance at 280 nm to be sure you have cleared out all the bacterial proteins that do not bind galactose. A_{280} must be 0.003 or less.

(8) Mark 15–20 2 ml microcentrifuge tubes to indicate 1.5 ml on the outside, using a marking pen. Label them 'LT eluent'.

Place your collection tubes in a rack, ready to go. *This is the elution of LTB – Don't miss a drop!*

(9) Apply 2 ml of 0.3 M galactose in TEAN. Immediately begin collecting 1.5 ml aliquots. Check the absorbance of the eluent fractions at 280 nm using an ethanol washed and air-dried 1 ml cuvette. It should be significantly higher than the last wash sample, and it should increase sharply. Return aliquots to their collection tube.

(10) Continue to add 0.3 M galactose and to collect 1.5 ml fractions, checking each absorbance. Absorbance should peak, then come down to near 0. When it comes near baseline (0.02 or less), all the LT has been eluted.

(11) To be certain the column is cleared, do a final wash with 1–2 ml of 1 M galactose, and check absorbance to verify that no additional protein elutes.

(12) Save all the tubes in a tray, labeled. Store in the refrigerator. Do not freeze – the TEAN will keep them growth-free. Multimeric proteins like LTB or CTB tend to aggregate and even denature when frozen and thawed.

(13) Wash the column with 10 volumes or more of TEAN. Since the buffer has EDTA and azide in it, the resin, packed, or unpacked and pooled, can be safely stored at 4°C for re-use. *Never freeze the resin* – that will crush the beads and make it unusable.

Preparation for experiment

(1) TEAN buffer: 50 mM Tris–HCl/pH 7.4, containing 1 mM EDTA, 3 mM NaN_3, 200 mM NaCl (see Experiment 18).

(2) 1 M D(+)-galactose in TEAN buffer (see above).

(3) 0.3 M D(+)-galactose in TEAN Buffer (see above).

References

Dallas, W. S. and Falkow, S. (1980). Amino acid sequence homology between cholera toxin and *Escherichia coli* heat-labile toxin. *Nature (London)* **288**, 499–501.

Merritt, E. A., Sixma, T. K., Kalk, K. H., Zanten, B. A. M. v. and Hol, W. G. J. (1994). Galactose binding site in *Escherichia coli heat-labile* enterotoxin (LT) and cholera toxin (CT). *Mol. Microbiol.* **13**, 745–753.

Tsuji, T., Honda, T., Miwatani, T., Wakabayashi, S. and Matsubara, H. (1985). Analysis of receptor-binding site in *Escherichia coli* enterotoxin. *J. Biol. Chem.* **260**, 8552–8558.

Uesaka, Y., Otsuka, Y., Lin, Z., Yamasaki, S., Yamaoka, J., Kurazono, H. and Takeda, Y. (1994). Simple method of purification of *Escherichia coli* heat-labile enterotoxin and cholera toxin using immobilized galactose. *Microbial Pathogen.* **16**, 71–76.

EXPERIMENT 20

(a) SDS–PAGE of Column Fractions
(b) Immunoblot to Verify Presence of LTB

Introduction

Protein analysis

The 280 nm absorbance tells absolutely nothing about protein size, purity (presence or absence of contaminating protein), aggregation state, activity or structure, let alone specific identity of the material in the column fraction. SDS–PAGE will provide a clear picture of the purity, the molecular weight of the sample and whether it is monomeric or pentameric, but cannot tell what the protein is, specifically, even though the molecular weight of an unheated vs a heated fraction can give a reasonable indication that it is the target pentamer. An immunoblot will identify the presence of a specific protein, but gives no information about purity, size or aggregation state. Importantly, neither technique can demonstrate whether the protein is active or if it is folded in the correct three-dimensional structure. To obtain that information activity assays are necessary, specifically, binding of LTB to ganglioside receptor analogs. X-ray structure determination would verify the tertiary structure. For that purpose, the isolated and purified protein must be crystallized.

Samples of column fractions chosen from the flow-through and the up-swing, peak and down-slope of the elution peaks (280 nm) will be checked to get an idea of what proteins are in the fractions. Based on the results, fractions can be pooled if they are identical single bands with no contaminating bands. A single pure fraction can be used if it contains enough material. The electrophoresis will be done in the same way as in Experiment 16. The precast gels are 4–12% polyacrylamide, cast in a gradient from top to bottom. This type of formulation provides very clean band separation. Heated and unheated samples of the peak fraction will be electrophoresed. An affinity column has been used to isolate a molecule that binds galactose, as pentameric LTB is expected to do. It is now necessary to verify that the molecular weight of the unboiled sample is approximately 60 000 Da and the boiled sample is approximately 12 000 Da by SDS—PAGE. This pattern is characteristic of LTB (and CTB) and could be considered diagnostic.

To be absolutely certain the protein is LTB, however, an immunoblot (ELISA) using rabbit anti-LT antibody that has been absorbed with *E. coli* lysate may be used to confirm the presence of LTB. Treating the primary antibody with *E. coli* lysate is well-advised for this particular experiment because the antigen used to immunize rabbits is often simply a lysate from pathogenic *E. coli* carrying the LT plasmid and expressing LT. In that case, the rabbit will produce antibody against LT, and against various *E. coli* proteins as well. The immobilized *E. coli* lysate affinity chromatography column removes any IgG elicited against *E. coli* proteins. The *E. coli* lysate affinity chromatography column is run the same way as the galactose affinity chromatography column. However, the flow-through is collected and saved, rather than the bound material, since it should contain anti-LT IgG, while the adsorbed material remaining on the column should be anti-*E. coli* IgGs. The purified anti-LT IgG will not react with any *E. coli* proteins that may have contaminated the galactose-binding protein previously isolated, and a false positive reaction is avoided. In addition, the antibody should not react with flow-though fractions that contain all *E. coli* proteins except LT.

However, detection with anti-CTB instead of anti-LT has several advantages: anti-CTB cross-reacts very well with LTB protein (B. D. Spangler, unpublished data); it is elicited in goats by challenge with *Vibrio cholerae*-derived purified CTB protein, therefore the goat serum will not contain any antibodies directed against *E. coli* that might provide a false positive background; and it is commercially available.

Materials and Methods

Materials

Electrophoresis

♦ Vertical polyacrylamide minigel electrophoresis apparatus (gel size 8.6 × 6.8 cm × 1 mm)

- Power supply

- P-200 Pipetman, gel-loading tips

- Precast 4–20% acrylamide Tris–HCl gels

- Unstained or pre-stained broad-range molecular weight markers, 200 000–7000

- 1× Tris–glycine/SDS running buffer (5 mM Tris/200 mM glycine/0.02% SDS, pH ~8.3; diluted from 5× buffer)

- 6× or 10× SDS–PAGE sample loading buffer (SDS–PAGE-LB)

- Coomassie Brilliant Blue stain (0.25% in 50% methanol/10% acetic acid solution)

- 15% glacial acetic acid or 30% methanol/10% acetic acid for destain

Immunoblot

- Column fractions from Experiment 19

- Immobilon-P transfer membrane (Millipore Corp., Bedford, MA, USA, catalog no. IPUH 101 00)

- 100% methanol

- Filter paper

- Transfer buffer (25 mM Tris base, 192 mM glycine, 10% v/v methanol)

- 1:1000 anti-CTB (goat anti-choleragenoid, List Laboratories, Campbell, CA, USA, catalog no. 703) diluted from a 1:50 stock

- Wash buffer (0.1% BSA/PBS, pH 7.2 containing 0.05% Tween-20)

- 1:1000 anti-goat IgG-HRP diluted from a 1:50 stock

- TMB blotting or CN naphthol or Fast-DAB precipitating chromophore

- Lysozyme or denatured *E. coli* cells as negative control

- Commercially obtained LT, CT or CTB as positive control, diluted to a final concentration of 10 μg/ml (List Laboratories, Campbell, CA, USA, catalog no 101B, C, D)

Method

Electrophoresis (SDS–PAGE)

The directions given below are for the BioRad Ready Gel Electrophoresis Cell (catalog no. 165-3125 or 165-3126).

Assembly

(1) Remove the electrode assembly form the clamping frame. Rotate the cams outward to release the electrode assembly.

(2) Prepare a precast Ready Gel by removing the tab from the bottom of the cassette. Repeat for a second gel. If only one gel is run, place the buffer dam in the empty slot.

(3) Place the cassettes into the slots at the bottom of each side of the electrode assembly. Be sure that the short glass plate of the Ready Gel cassette faces *inward* toward the notches on the green U-shaped gaskets.

(4) Press the Ready Gel cassette up against the gaskets. These fit together to create the inner chamber.

(5) Transfer the electrode assembly and gels into the clamping frame.

(6) Press down on the electrode assembly with forefingers of each hand while simultaneously closing the two cam levers of the clamping frame with the thumbs of each hand.

(7) Lower the electrode assembly and clamping frame into the minitank.

(8) Fill the inner chamber with 125 ml of running buffer, so that the buffer reaches a level between the tops of the short and long plates of the Ready Gel. Do not overfill.

(9) Add 200 ml of running buffer to the minitank.

Sample preparation

(1) Remove two 15 μl aliquots from a flow-through sample and place each in a 1.5 or 0.5 ml microcentrifuge tube. Remove two 20 μl aliquots each from the first eluent tube, from the peak, and from the downslope eluent samples. Label the tubes.

(2) Add 3 μl 6× or 2 μl 10× SDS—PAGE sample buffer (SDS—PAGE LB) to each flow-through tube. Add 4 μl 6× or 3 μl 10× SDS—PAGE sample buffer (SDS—PAGE LB) to each eluent tube. Mix by vortexing. Replace the column fractions in the refrigerator.

(3) Label one tube of each pair 'heated' and the other 'unheated'.

(4) Leave the tubes labeled 'unheated' at room temperature. Heat the other tubes to 90°C for 4 min in a beaker filled with water to dissociate all oligomeric proteins.

(5) Load 20–25 μl of each sample/well, precisely noting the contents of each well.

(6) Load one lane of each gel with molecular weight (MW) markers.

(7) Electrophorese for ~45 min, at 150–200 V (60 mA current)/gel until the tracking dye is 1–2 cm from the bottom.

(8) Turn off the current, unplug the power supply, then remove the gel using gloves so you do not leave protein fingerprints on the gel – they will stain nicely with Coomassie Blue!

(9) Transfer the gel to a container filled with Coomassie stain. Stain overnight or use a zinc stain kit according to manufacturer's instructions.

(10) Destain for at least 24 h in destain solution.

The gel may be preserved by subsequent soaking for several hours in destain containing 15% glycerol, then drained and placed between two heavy transparent sheets. Weighting it with a flat object will help keep it from curling while it dries out. They can be preserved as an electronic image or duplicated on a copy machine.

Immunoblot (while gel is running)

Immobilon-P transfer membranes can be used, omitting some traditional immunoblot steps. Unlike nitrocellulose, Immobilon-P does not bind proteins once it has been dried, which minimizes or eliminates a preliminary blocking step.

Use forceps to handle the membrane. Wear gloves to avoid getting fingerprints on the membrane.

(1) Draw 2 mm circles on a strip of dry Immobilon-P. Mark a number under the circles. Use a soft pencil, not ink – the membranes will be washed in methanol.

(2) Using forceps, place membrane in a Petri dish or a 10 ml test tube containing 100% methanol for 10 s.

(3) Shift the membrane to another test tube containing distilled water and soak for 2 min.

(4) Shift the membrane to the transfer buffer and soak for 5 min.

(5) Wet a piece of filter paper with transfer buffer and place it in a small box or Petri dish lid.

(6) Shift the Immobilon-P to the moistened filter paper and cover with a lid to prevent the membrane from drying out.

(7) Load 5 µl of 'pre-column' into circle 1; load 5 µl of 'flow-through' into circle 2; load 5 µl of galactose eluate into circle 3; load 5 µl of negative control into circle 4; load 5 µl of positive control (LT) into circle 5.

(8) Allow the samples to soak into the membrane for ~5 min.

(9) Remove the membrane from the filter paper and place in methanol for 10 s.

(10) Place the membrane on a piece of dry filter paper for ~15 min or place in a warm oven to dry thoroughly. These steps will drive out the water and make the membrane too hydrophobic to absorb any protein, including nonspecific adsorption of antibodies. If the membrane is not entirely dry, the background will be dark and any positive results will be invisible.

(11) Place the strip(s) in a test tube containing 5 ml 1:1000 anti-CTB primary antibody. Place the tube in a beaker and gently rock for 45 min.

(12) Remove the strips from primary antibody solution and place in a tube of wash buffer (PBS or 1% BSA/PBS/Tween).

(13) Pour out the wash buffer and add fresh buffer. Repeat once more. This constitutes 'washing three times in PBS'.

(14) Place strips in a tube containing 5 ml 1:1000 anti-goat IgG-HRP (secondary antibody) and incubate for 30 min with rocking.

(15) Wash the strips three times in wash buffer [see steps (12)–(13)].

(16) Place the strips in a tube containing TMB (blotting) or other precipitating chromophore and incubate for 5–10 min.

(17) Remove the strips and lay them flat on a piece of filter paper. Record the positions and identities of the colored spots. Note the intensities using +, ++, +++ to evaluate color development.

Preparation for experiment

(1) 2× SDS–PAGE sample loading buffer (Sambrook and Russell, 2000): 100 mM Tris–HCl, pH 6.8 (see Experiment 2 for Tris buffer preparation); 200 mM DTT added just before use*; 4% (w/v) electrophoresis grade SDS; 0.2% bromophenol blue; and 20 % (v/v) glycerol. Store aliquots frozen.

---

*Store 2× SDS-PAGE LB without DTT at room temperature. Add DTT from a 1 M stock just before the buffer is to be used or add 0.154 g directly to 5 ml prepared loading buffer. Several microliters of 2-mercaptoethanol may be added directly to the prepared 2× buffer in place of the DTT immediately prior to use. Sample buffer, 6× or 10\times can be prepared from suitably adjusted amounts of reagents, or can be purchased commercially.

(2) Coomassie Brilliant Blue stain (0.25% in 50% methanol/10% acetic acid solution; see Experiment 16).

(3) Transfer buffer 25 mM Tris base, 192 mM glycine, 10% v/v methanol (see Experiment 16).

(4) Wash buffer (0.1% BSA/PBS, pH 7.2 containing 0.05% Tween-20): 99 ml PBS, pH 7.2 (see Experiment 18); 1 ml 10% BSA (see Experiment 18); 50 μl Tween-20.

(5) Anti-goat IgG-HRP (secondary Ab) 1:50 stock in 1% BSA/PBS or wash buffer: 100 μl anti-goat IgG-HRP; 5 μl PBS/1% BSA (0.5 μl 10% BSA, 4.5 ml PBS).

(6) TMB blotting or CN naphthol or fast DAB precipitating chromophore (see Experiment 17).

(7) 1:1000 dilution: 250 μl 1:50 stock in 5 ml PBS.

Note: TBS may be substituted for PBS (see Experiment 17).

Reference

Sambrook, J. and Russell, D. W. (2000). *Molecular Cloning: a Laboratory Manual*. Cold Spring Harbor, New York: Cold Spring Harbor Laboratory Press.

EXPERIMENT *21*

Protein Concentration and Protein Crystallization

Introduction

Understanding the structure of a protein is crucial to determining how it functions. X-ray crystallography allows the three-dimensional molecular structure of proteins and other macromolecules having molecular weights (MW) 10 000–200 000 Da to be determined. For larger molecules the algorithms and data gathering is much more difficult, unless the molecule is composed of repeating subunits. For mid-sized molecules (MW 10 000–20 000) NMR is a very useful structural tool. For small molecules (MW 50–5000) X-ray crystallography is common practice. Knowledge of the three-dimensional structure allows structure–function inferences to be made and to design new drugs that may be used to manipulate the molecular function (see Experiment 24 for a more detailed discussion).

The LTB protein is pentameric, with a total molecular weight of ~60 000. Although its structure has been solved previously (Sixma *et al.*, 1991), and there are structures showing the LTB bound to various ligands (Merritt *et al.*, 1994; Sixma *et al.*, 1992), a crystallographic dataset will be required for comparison with any mutated proteins produced later.

For X-ray analysis, it is necessary to obtain crystals of molecules because a crystal contains many hundreds of molecules in a repeating pattern. Such repetitive positioning of the atoms enhances (reinforces) the intensity of the data and makes measurement of atomic positions possible. Once a detailed picture of the contents of a crystal has been obtained and the positions of the individual atoms are known, one can calculate the interatomic distances, bond angles and other features of molecular architecture. Several texts are available and are listed at the end of Experiment 24.

The largest hurdle to be overcome when solving a structure is to obtain crystals of high enough quality, or large enough size, for crystallographic studies. Obviously the larger the crystal, the more molecules are present, and the stronger the data that can be collected. Generally, protein crystals have dimensions of about 0.3 mm–1.0 mm and the molecules must be arranged in an orderly array. To induce crystallization, a protein solution of high concentration (10–50 mg/ml) is brought nearly to spontaneous precipitation by the addition of a precipitating reagent. Then, by slowly increasing the concentration of protein and precipitant, one hopes that crystals will form in an orderly way and continue to grow. Often, many variations must be tried to find conditions that will produce well-ordered, good-sized, crystals. For example, LT can crystallize when stored in the refrigerator in buffer. The crystals are not, however, suitable for X-ray diffraction studies because they are disordered. Pronk *et al.* (1985) tried as many as 2000 different conditions before obtaining crystals suitable for diffraction studies. Her conditions and some simpler versions have been used in several recent studies, including the Merritt article cited above.

Note especially that LT is practically insoluble at low ionic strength, so buffers must contain sufficient salt (or high enough molarity) to maintain solubility, otherwise it cannot be concentrated. It will aggregate and precipitate.

Most crystallization recipes are determined by trial and error, although starting conditions are often guessed based on conditions that approach but do not mimic those that cause aggregation or precipitation of the protein. Dilute protein solutions at the isoelectric pH of the protein, or using organic precipitants that, in higher concentrations, cause insolubility are common starting points. A matrix plate, with 8–24 different conditions, can be obtained by varying several parameters (pH, salt type and concentration, buffer ionic strength, precipitant, for example). Pre-mixed reagents for matrix plates and other tools for protein crystallization can be obtained from Hampton Research Inc. (Riverside, CA, USA; www.hamptonresearch.com).

Pronk *et al.* (1985) devised a micro-dialysis technique to crystallize LT, which was useful because it allowed crystals that formed to be easily manipulated. Although it is still used, the hanging drop-vapor diffusion method and the sitting drop method are most frequently employed now. In the hanging drop vapor diffusion method, the protein solution is diluted 1:1 with reservoir buffer so the drop contains half the solute concentration compared to the reservoir and half the starting protein concentration. Therefore, vapor diffuses from the drop to the reservoir, leading to very gradual dehydration of the drop, concentrating protein and solute.

Lysozyme, an enzyme found in egg whites, crystallizes into several crystal forms, and does so very easily, even when egg white is treated with sodium hydroxide and NaCl, or with vinegar and salt. Several crystallization recipes for lysozyme will be compared as a demonstration.

Materials and Methods

Materials

♦ Inverted dissecting microscope for viewing crystals

♦ Microconcentrators, 30 000 molecular weight cut-off (MWCO)

♦ Polyethylene glycol average MW 8000 (formerly listed as PEG 6000, Sigma-Aldrich, St Louis, MO, USA catalog no. P 2139 or Hampton Research, Riverside, CA, USA)

♦ 24-well sterile culture plates with lids, Linbro boxes, or VDX plates (Hampton Research, Riverside CA, USA catalog no. HR3-114; HR3-110; HR3-142)

♦ Microbridges for sitting drops in Linbro or VDX plates (Hampton Research, Riverside, CA, USA catalog no. HR3-310); alternatively, a nine-well depression plate on plastic Petri dish bottom, set in a plastic sandwich box (Hampton Research, Riverside CA, USA, catalog no. HR3-130; HR3-132)

♦ 10 ml disposable syringe, filled with vacuum grease

♦ Siliconized cover slides (Hampton Research, Riverside CA, USA, catalog no. HR3-231 or HR3-215)

♦ 100 mM Tris Buffer, 200 mM NaCl, pH 7.4 for concentration and storage of LTB

♦ 15% PEG 8000 in 100 mM Tris, 50 mM NaCl, pH 7.4

♦ 30% PEG 8000 in 100 mM Tris, 50 mM NaCl, pH 7.4

♦ rLTB (recombinant LTB protein) from Experiment 20, previously dialyzed or buffer-exchanged with 100 mM Tris, 200 mM NaCl, pH 7.4, to remove galactose.

For lysozyme crystallization (conditions modified from McPherson, 1999):

♦ 20 mg/ml lysozyme in distilled water (Sigma-Aldrich, St Louis, MO, USA catalog no. L-7773 only)

♦ Solution A: 0.1 M sodium acetate, pH 4.5, 0.5 M NaCl

♦ Solution B: 0.1 M sodium acetate, pH 4.7, 0.5 M NaCl

♦ Solution C: 0.1 M sodium acetate, pH 4.5, 1.0 M NaCl

♦ Solution D: 0.1 M sodium acetate, pH 4.7, 1.0 M NaCl

Method: rapid concentration and buffer exchange of rLTB

(1) Pool pure identical eluent fractions of rLTB (single band, no contaminants as determined by SDS–PAGE and immunoblot, Experiment 20)

(2)   Estimate rLTB protein concentration spectrophotometrically. The molar absorptivity (ε_{280}) of CTB is 1.43×10^4/M/cm and the $A_{1\%}$ at 280 nm [the absorption of a 1% (1g/100 ml) solution] is 9.56 (Finkelstein, 1973). The numbers for LTB are very similar. Using the molar absorptivity and Beer's law, calculate moles and convert to mg/ml for this experiment, assuming a molecular weight of approximately 59 000 Da for the LTB pentamer. Alternatively, set up a ratio equation using the measured A_{280} and the $A_{1\%}$ and calculate mg/ml directly.

(2)   Place 0.5–1 ml pooled protein solution in a 30 000 MWCO microconcentrator. Centrifuge for 10 min at 7000–10 000 g. The solution should be reduced to approximately 50–100 µl.

(3)   Add 1 ml of 100 mM Tris, 200 mM NaCl, pH 7.4 (a 10–20 vol excess) and re-concentrate to the 50–100 µl in the microconcentrator. Centrifuge for 10 min at 7000–10 000 g. Repeat one more time. This procedure will effectively exchange the buffer and remove the galactose while concentrating the protein. Crystallization requires 4–6 mg/ml in 100 mM Tris, 200 mM NaCl, pH 7.4. A higher concentration, up to 10 mg/ml is even better.

Crystallization of rLTB

This protocol was suggested by Professor Wim Hol and Ms Misol Ahn (University of Washington, Seattle, USA, personal communication).

Note: if there is insufficient material, do only one row for each condition, three wells hanging drop, three wells sitting drop. More than one condition will be tried to determine which works best for this particular preparation.

(1)   Add 1 ml 15% PEG 8000 reservoir solution to wells, A, B (1–6)

(2)   Add 1 ml 30% PEG 8000 reservoir solution to wells C, D (1–6)

(3)   Place a microbridge into each well A, B, C, D (4–6)

(4)   Place a bead of vacuum grease around the edge of each well, expressing it carefully from the syringe. Do not be too generous, but the well must be rimmed sufficiently to ensure a good seal between coverslip and rim.

(5)   Set coverslips on the plate lid so the edges are just hanging over the edge of the lid (to facilitate picking them up after loading).

(6)   Place 5 µl concentrated protein solution on three coverslips at a time. Add 5 µl reservoir solution taken from the well over which the drop will hang. *Do not mix by*

pipetting up and down. Mixing, stirring or scratching will result in a shower of small crystals.

(7) Carefully pick up each coverslip and invert, then place gently on a beaded well (A, B, 1–3). Press *very* gently to seal the coverslip over the well.

(8) For the microbridge wells, add 5 µl reservoir buffer from the well to the microbridge seat, and add 5 µl concentrated rLTB to that.

(9) Grease around the top edge of the well and place a coverslip carefully but firmly on top.

Put lids on plate and store on a shelf and *do not disturb* for at least 3–5 days. Crystals can be observed with an inverted dissecting microscope and indirect light.

Crystallization of lysozyme (20 mg/ml stock)

(1) Set up one plate:
   1 ml/well: A(1–6) Solution A (1–3 for hanging drops; 4–6 with microbridge sitting drops*);
   B(1–6) Solution B (1–3 for hanging drops; 4–6 with microbridge sitting drops);
   C(1–6) Solution C (1–3 for hanging drops; 4–6 with microbridge sitting drops);
   D(1–6) Solution D (1–3 for hanging drops; 4–6 with microbridge sitting drops).

(2) Place a bead of vacuum grease around the edge of each well, expressing it carefully from the syringe. Do not be too generous, but the well must be rimmed sufficiently to ensure a good seal between coverslip and rim.

(3) Set coverslips on the plate lid so the edges are just hanging over the edge of the lid (to facilitate picking them up after loading).

(4) Place 5 µl lysozyme (20 mg/ml) carefully in the center of three coverslips. Add 5 µl reservoir solution taken from wells A(1–3) to each drop. *Do not mix by pipetting up and down.* Mixing, stirring, or scratching will result in a shower of small crystals.

(5) Carefully pick up each coverslip and invert, then place gently on a beaded well (A1–3). Press *very* gently to seal the coverslip over the well.

(6) Repeat in triplicate for B (1–3); C (1–3); D (1–3).

---

*See below for sitting drop alternative.

(7) For microbridges, place 5 µl reservoir and 5 µl lysozyme into each depression. Grease the top edge of the well and place a coverslip carefully but firmly on top of the wells containing filled microbridges.

Alternatively, a clear plastic sandwich box, containing a square or round Petri dish bottom set upside down and topped with a nine-well depression plate can be used. Place the reservoir solution in the bottom of the sandwich box, one buffer formulation per box. Add 15 µl reservoir buffer to each well and carefully add 15 µl of 20 mg/ml lysozyme stock. *Do not mix!* Bead the edge of the sandwich box with vacuum grease or Vaseline and place the lid on tightly. Wrap the lid to the box with Parafilm to prevent evaporation of the reservoir and drops.

Lysozyme, apparently, can be crystallized out of egg white in the following way:

(a) separate an egg, discarding the yolk (lysozyme is the major protein in egg whites);

(b) place the egg white in a 10 ml beaker or vial; estimate its volume and add a small stir bar;

(c) adjust the pH to 10.5 with 1 M NaOH stirring gently; check the pH with pH paper;

(d) add NaCl to a final concentration of 5% (50 mg/ml), stirring gently;

(e) drop in a few seed crystals (pipet from a P-1000);

(f) cover the beaker with Parafilm (or cap the vial);

(g) Let it sit. Crystals may form in days (or a week).

Dr Enrico Stura (CEA Saclay, Gif sur Yvette, France) suggests a rapid method (crystals in 15 min): 100 mg/ml lysozyme in 50 mM sodium acetate, pH 4.5, and a reservoir solution containing 30% w/v PEG 5000, 1.0 M NaCl, 50 mM sodium acetate, pH 4.5. The protein solution is diluted 1:1 in the reservoir solution. A hanging drop or sitting drops can be used. To obtain larger crystals, dilute the lysozyme or use less PEG. This method can be done at 4°C or 22°C.

Preparation for experiment

(1) 100 mM Tris–HCl, pH 7.4/200 mM NaCl.

(2) 100 mM Tris, 50 mM NaCl, pH 7.4: 1 ml 1 M Tris stock; 8 ml distilled water. Adjust pH to 7.4 with concentrated HCl. Add 0.029 g NaCl. Make up to 10 ml.

(3) 15% PEG 8000 in 100 mM Tris, 50 mM NaCl, pH 7.4; 1.5 g PEG 8000; 8.5 ml 100 mM Tris, 50 mM NaCl, pH 7.4.

(4) 30% PEG 8000 in 100 mM Tris, 50 mM NaCl, pH 7.4: 3 g PEG 8000; 7 ml 100 mM Tris, 50 mM NaCl, pH 7.4

(5) rLTB from Experiment 20, previously dialyzed against 100 mM Tris, 200 mM NaCl, pH 7.4 to remove galactose or buffer exchanged/concentrated see above.

(6) For lysozyme crystallization: 20 mg/ml lysozyme in distilled water (lysozyme Sigma-Aldrich catalog no. 7773 will crystallize very readily; it may be possible to substitute catalog no. 6876 if the solution is 0.2 μ filtered before use);

0.1 M sodium acetate, pH 4.5 [575 μl (0.575 ml) glacial acetic acid; 90 ml distilled water]. Adjust 9 ml to pH 4.5 with 1 M NaOH. Make up to 10 ml. Adjust another 9 ml to pH 4.7 with 1 M NaOH. Make up to 10 ml.

(a) Solution A: 0.1 M sodium acetate, pH 4.5, 0.5 M NaCl (0.29 g NaCl; 10 ml 0.1 M sodium acetate, pH 4.5).

(b) Solution B: 0.1 M sodium acetate, pH 4.7, 0.5 M NaCl (0.29 g NaCl; 10 ml 0.1 M sodium acetate, pH 4.7).

(c) Solution C: 0.1 M sodium acetate, pH 4.5, 1.0 M NaCl (0.58 g NaCl; 10 ml 0.1 M sodium acetate, pH 4.5).

(d) Solution D: 0.1 M sodium acetate, pH 4.7, 1.0 M NaCl (0.58 g NaCl; 10 ml 0.1 M sodium acetate, pH 4.7).

The solutions can also be prepared from 1 M sodium acetate buffer, pH suitably adjusted, and 2 M NaCl in water.

References

Finkelstein, R. A. (1973). Cholera. *Crit. Rev. Microbiol.* **2**, 553–623.

McPherson, A (1999). *Crystallization of Biological Macromolecules*. Cold Spring Harbor Press, NY.

Merritt, E. A., Sixma, T. K., Kalk, K. H., Zanten, B. A. M. v. and Hol, W. G. J. (1994). Galactose binding site in *Escherichia coli* heat-labile enterotoxin (LT) and cholera toxin (CT). *Mol. Microbiol.* **13**, 745–753.

Pronk, S. E., Hofstra, H., Groendijk, H., Kingma, J., Swarte, M. B. A., F. Dorner, J. D., Hol, W. G. J. and Witholt, B. (1985). Heat-labile enterotoxin of *E. coli:* characterization of different crystal forms. *J. Biol. Chem.* **260**, 13580–13584.

Sixma, T. K., Pronk, S. E., Kalk, K. H., Wartna, E. S., Zanten, B. A. M. v., Witholt, B. and Hol, W. G. J. (1991). Crystal structure of a cholera toxin-related heat-labile enterotoxin from *E. coli*. *Nature (Lond.)* **351**, 371–377.

Sixma, T. K., Pronk, S. E., Kalk, K. H., Zanten, B. A. M. v., Berghuis, A. M. and Hol, W. G. J. (1992). Lactose binding to heat-labile enterotoxin revealed by X-ray crystallography. *Nature (Lond.)* **355**, 561–564.

EXPERIMENT *22*

Setting up an ELISA Microtiter Plate to Assay a Panel of Ganglioside Ligands

Introduction

ELISA is an immunochemical method (meaning it uses antigen–antibody interactions) that is suitable for qualitative detection of specific antigens. With proper controls and standard curves, it can be semi-quantitative as well. The method was introduced in 1971 and has been adopted as an industry standard practice for assay of all types of molecules for which antibodies can be made. The basic method as applied to a two-dimensional blot was described in Experiments 17 and 20. In this experiment, the assay is done in a 96-well microtiter plate. There are several ways of performing the assay, depending on what is to be measured:

(1) In the original method, the antibody sandwich assay, the antigen is 'sandwiched' between adsorbed antibody and conjugated antibody. Antibodies against (elicited by) the antigen to be measured are adsorbed on a solid support, usually a 96-well polystyrene microtiter plate. After coating the well bottom with antibody by adsorption from solution, then washing off any unadsorbed material, the well and sample are treated with a non-specific protein, such as BSA or milk proteins, to block any non-specific binding to the antibody or to the walls of the well. Antigen is added, and will bind to the antibody molecules adsorbed on the bottom of the well. Next, a conjugate that will also bind to the antigen is added. The conjugates are antibodies bound to an

enzyme. After addition of a chromogenic (color-producing) substrate for the enzyme, the intensity of the colored product generated will be proportional to the amount of conjugated enzyme and, indirectly, to the amount of antigen bound by the adsorbed antibody.

(2) Antibody titers can be compared by adsorbing antigen to the microtiter plate wells then adding serial dilutions of antibody. The bound antibody is probed with anti-immunoglobulin-specific antibody conjugated to enzyme, then quantitated with chromogenic substrate.

(3) Antigen-capture assays make use of a capture ligand, for example a receptor analog, coated on the sides of the wells. The antigen is serially diluted and added to the wells. The amount of antigen bound by the capture agent is assayed by addition of specific antibody against the antigen, followed by anti-immunoglobulin conjugate, then chromogenic substrate.

(4) Competition assays attempt to determine whether the antibody–antigen complex on the plate can be selectively displaced by added antigen. In some cases, a heterogeneous solution containing antigen is adsorbed on the plate and antibody added. Then a known concentration of pure antigen is added. The pure antigen should bind to antibody, displacing it from adsorbed antigen. In that case, the antibody remaining is probed with anti-immunoglobulin conjugate and chromogenic substrate. The amount of antibody bound is directly proportional to the amount of antigen in the well. Alternatively, the amount of antigen in the heterogeneous solution can be estimated by using serial dilutions of antigen in solution. The remaining antibody is quantitated using anti-immunoglobulin antibody conjugated to enzyme, then chromogenic substrate.

The different types of commonly used ELISA are diagrammed below (Figure 22.1). There are several other variations for particular purposes, as well.

In the next experiment the antigen-capture ELISA will be used to evaluate the amount of LTB bound to several different ganglioside receptor analogs. In this experiment, the microtiter plate will be coated with gangliosides. For this purpose poly(vinyl chloride) (PVC) microtiter plates are used rather than the polystyrene plates used for proteinaceous capture agents, because the hydrophobic long chain alkyl groups on gangliosides do not adsorb to polystyrene. U-bottom PVC microtiter plates are used rather than flat-bottom plates to conserve reagents. U-bottom or modified V-bottom plates require only 50 µl of sample to cover the bottom, while flat-bottom plates require 100 µl of sample. In addition, the U-bottom or V-bottom wells can be washed clear of unbound material more effectively than flat-bottom wells.

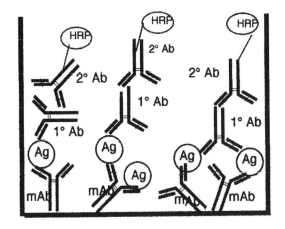

Sandwich ELISA

Ab = antibody

mAb = monoclonal antibody

Ag = antigen

HRP = horseradish peroxidase

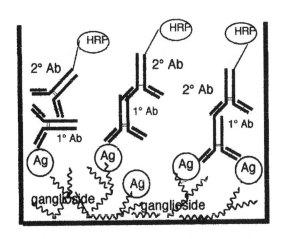

Antigen-capture ELISA

on a receptor-analog-coated well

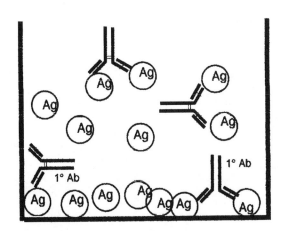

Competition ELISA

soluble antigen competes with
bound antigen for antibody

Figure 22.1 Microtiter plate techniques

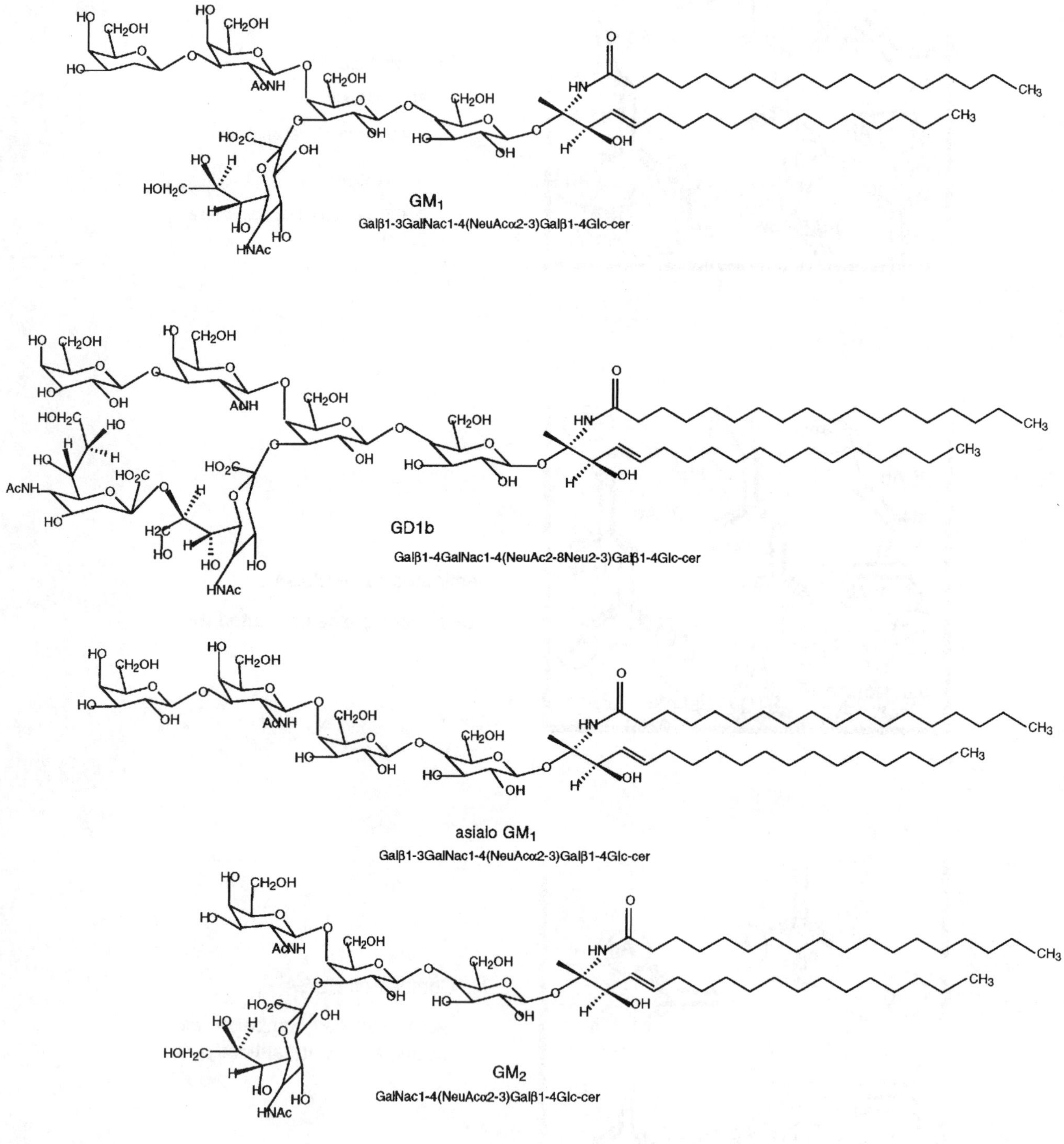

GM₁

Galβ1-3GalNac1-4(NeuAcα2-3)Galβ1-4Glc-cer

GD1b

Galβ1-4GalNac1-4(NeuAc2-8Neu2-3)Galβ1-4Glc-cer

asialo GM₁

Galβ1-3GalNac1-4(NeuAcα2-3)Galβ1-4Glc-cer

GM₂

GalNac1-4(NeuAcα2-3)Galβ1-4Glc-cer

Figure 22.2   Structures of some ganglioside receptor analogs

Materials and Methods

Materials

♦ 96-well U-bottom PVC microtiter plates (Becton-Dickenson)

♦ P-200 Pipetmen, white or yellow tips

♦ Disposable sealing tape

♦ Ganglioside GM_1 stock 20 μg GM_1/ml methanol; for use, dilute 1:100 in methanol, so that 50 μl contains 10 ng GM_1

♦ Ganglioside GD_{1b} (20 μg GD_{1b}/ml methanol stock); for use, dilute 1:100 in methanol, so that 50 μl contains 10 ng

♦ Ganglioside GM_2 (20 μg GM_2/ml methanol stock); for use, dilute 1:100 in methanol, so that 50 μl contains 10 ng

♦ Ganglioside asialo-GM_1 (20 μg asialo-GM_1/ml methanol stock); for use, dilute 1:100 in methanol, so that 50 μl contains 10 ng

Ganglioside GT_{1b} or others may be substituted for GM_2 or asialo-GM_1.

Method

See Figure 22.3 for summary.

(1) Add 50 μl (10 ng) ganglioside GM_1 to wells A–H (1–3).

(2) Add 50 μl (10 ng) ganglioside GD_{1b} to wells A–H (4–6).

(3) Add 50 μl (10 ng) ganglioside GM_2 to wells A–H (7–9).

(4) Add 50 μl (10 ng) ganglioside asialo-GM_1 to wells A–H (10–12).

(5) Allow to air-dry overnight without cover or dry, open, in a vacuum dessicator at room temperature. Plates can be dried in a vacuum oven at 40–50°C for 20 min.

(6) Cover with disposable tape the next morning and store on the bench top.

Preparation for experiment

(1) Ganglioside GM_1 stock 20 μg GM_1/ml methanol stored in glass, at −20°C; for use, dilute 1:100 in methanol, so that 50 μl contains 10 ng GM_1 (Sigma-Aldrich catalog no. G7641; Matreya Inc., Pleasant Gap, PA, USA catalog no. 1061)

	GM₁			GD₁ᵦ			GM₂			asialo-GM₁		
	1	2	3	4	5	6	7	8	9	10	11	12
A												
B												
C												
D												
E												
F												
G												
H												

Figure 22.3   Plate diagram

(2) Ganglioside GD_{1b} (20 μg GD_{1b}/ml methanol stock) stored in glass, at $-20°C$; for use, dilute 1:100 in methanol, so that 50 μl contains 10 ng (Sigma-Aldrich catalog no. G2392; Matreya catalog no. 1501).

(3) Ganglioside GM_2 (20 μg GM_2/ml methanol stock) stored in glass, at $-20°C$; for use, dilute 1:100 in methanol, so that 50 μl contains 10 ng (Sigma-Aldrich catalog no. G8397; Matreya catalog no. 1502).

(4) Ganglioside asialo-GM_1 (20 μg asialo-GM_1/ml methanol stock) stored in glass, at $-20°C$; for use, dilute 1:100 in methanol, so that 50 μl contains 10 ng (Sigma-Aldrich catalog no. 3018; Matreya catalog no. 1064).

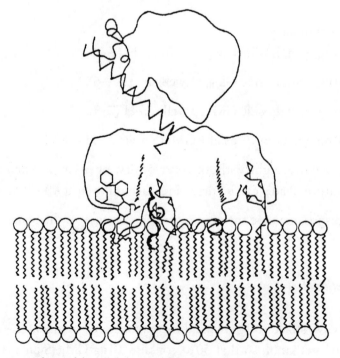

Figure 22.4   Cartoon of *E. coli* heat-labile enterotoxin associated with a cell membrane, Spangler, 1992. (Reproduced by permission of the American Society for Microbiology)

Other possibilities can be substituted, based on information found in Fukuta *et al.* (1988) and Merritt *et al.* (1995, 1997).

References

Fukuta, S., Twiddy, E. M., Magnani, J. L., Ginsburg, V. and Holmes, R. K. (1988). Comparison of the carbohydrate binding specificities of cholera toxin and *Escherichia coli* heat-labile enterotoxins LTH-I, LT-IIa, and LT-IIb. *Infect. Immun.* **56**, 1748–1753.

Merritt, E. A., Sarfaty, S., Chang, T.-t., Palmer, L. M., Jobling, M. G., Holmes, R. K. and Hol, W. G. J. (1995). Surprising leads for a cholera toxin receptor-binding antagonist: crystallographic studies of CTB mutants. *Structure* **3**, 561–570.

Merritt, E. A., Sarfaty, S., Jobling, M. G., Chang, T., Holmes, R. K., Hirst, T. R. and Hol, W. G. (1997). Structural studies of receptor binding by cholera toxin mutants. *Protein Sci.* **6**, 1516–1528.

Spangler, B. D. (1992). Structure and function of cholera toxin and the related *Escherichia coli* heat-labile enterotoxin. *Microbiol. Rev.* **56**, 622–647.

EXPERIMENT *23*

ELISA of a Panel of Ganglioside Ligands:

Introduction

Previous optimization experiments (E. A. Wilkinson and B. D. Spangler, unpublished results) have shown that:

(1) The optimal amount of ganglioside adsorbed in a microtiter plate well is between 1 and 10 ng GM1/well. Figure 23.1(a) shows that a titration of LT on 0.1 ng GM1/well results in a smaller net amount of LT bound than LT bound to 1 ng GM1/well, whereas, in a separate experiment, 1 ng GM1/well and 10 ng GM1/well are nearly the same [Figure 23.1(b)].

(2) Addition of more than 10 ng LT/well on 10 ng GM_1/well does not increase the amount of LT bound. That is, the concentration of LT required to saturate a well containing 1 ng adsorbed GM_1 or 10 ng adsorbed GM_1 is 1×10^5 pg LT/well (10 ng LT/well). While this is shown in Figure 23.2, the same result can be derived from the data shown in Figure 23.1.

(3) It has been determined that, given 10 ng adsorbed GM_1/well, and 10 ng LT/well, 5000 pg (5 ng) of rabbit anti-LT antibody produces maximum response using anti-rabbit IgG-peroxidase conjugate to assay bound anti-LT and TMB as the chromogenic substrate (Figure 23.3). Use of 1 mg LT/well does not change this amount (Figure 23.4), probably because 10 ng LT is the maximum amount of toxin that binds to capture ligand.

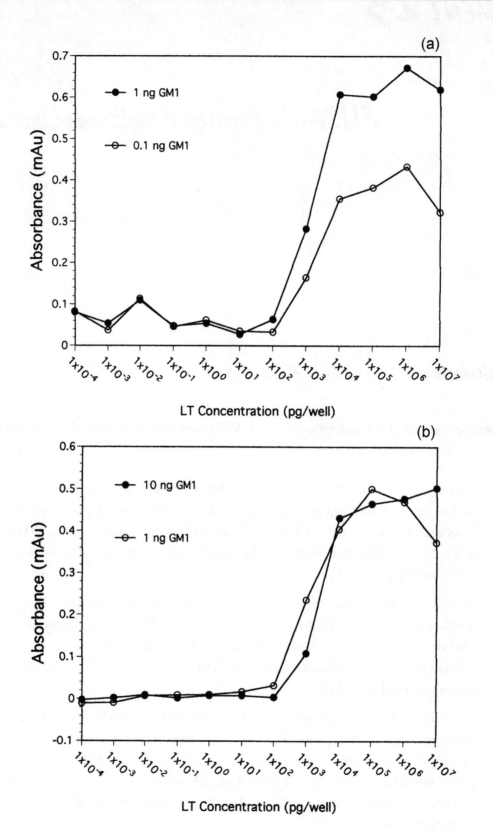

Figure 23.1   (a) Titration of LT on 0.1 ng and 1 ng GM₁/well. (b) Titration of LT on 1 and 10 ng GM₁/well

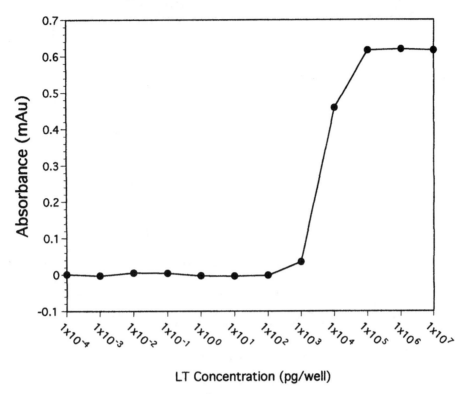

Figure 23.2 Titration of LT on 10 ng GM₁/well

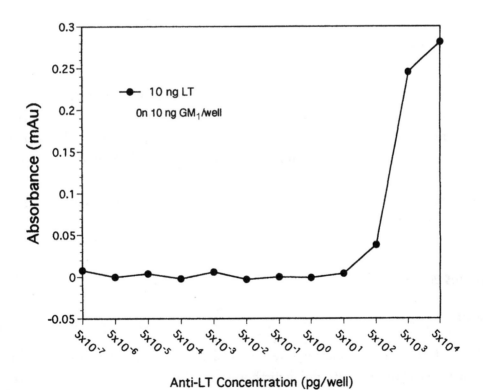

Figure 23.3 Titration of anti-LT IgG on 10 ng LT bound to 10 ng GM₁/well

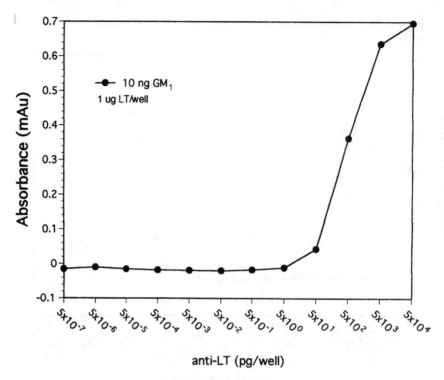

Figure 23.4   Titration of anti-LT IgG on $1 \, \mu$ LT bound to 10 ng Gm_1/well

(4) LT binding to GM_1 is virtually instantaneous [Figure 23.5(a)] while binding of anti-LT to immobilized LT is much slower, not reaching saturation until nearly 2400 s (40 min) after addition of antibody to the wells [Figure 23.5(b)].

Based on this data wells will be coated with 10 ng ganglioside capture agent/well. The assay for binding will use 10 ng LT/well. Anti-CTB will be used in this experiment.

LT binding will be evaluated using 10 ng LT/well in the upper half of the plate, while the lower half will be used to determine whether the cloned rLTB binds to the panel of gangliosides in a similar way.

Note: Experiments performed at different times or in different plates may not generate comparable absorbance readings.

Materials and Methods

Materials

♦ Microplate reader, if available (if not available, the color intensity can be estimated visually using the $+1$–$+10$ coding method)

♦ P-200 Pipetmen with tips

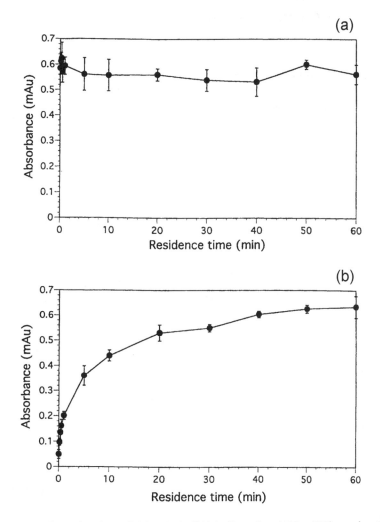

Figure 23.5 Time study: (a) binding of LT to GM$_1$; (b) binding of anti-LT to LT bound to GM$_1$ (reproduced by permission of Elsevier Science Ltd) (Spangler *et al.*, 2001)

◆ 1.5 ml microcentrifuge tubes

◆ 96-well PVC microtiter plate from Experiment 22

◆ PBS/1% BSA

◆ LT, CT or CTB (200 ng/ml PBS/1% BSA)

◆ LT, CT or CTB (100 ng/ml PBS/1% BSA)

◆ LT, CT or CTB (50 ng/ml PBS/1% BSA)

◆ rLTB (recombinant LTB) in 100 mM Tris, 200 mM NaCl, pH 7.4

◆ 1:5000 anti-CTB in PBS/1% BSA

♦ 1:5000 anti-goat IgG-HRP in PBS/1% BSA

♦ Wash buffer (PBS/0.1% BSA, pH 7.2 containing 0.05% Tween-20)

♦ TMB one-step or other chromophore/substrate (see Experiment 17)

♦ TBS may be substituted for PBS

Method

(1) Remove the tape covering the plate and allow the wells to air dry for 5 min to be sure all traces of methanol are gone.

(2) Blocking step: add 100 μl PBS/1% BSA (blocking buffer) to each well and incubate for 30 min, no longer, at room temperature. Longer blocking may lift some of the ganglioside coating.

(3) Prepare samples: serial two-fold dilutions. Label a set of 1.5 ml microcentrifuge tubes E, F, G.

(a) add 1.5 ml rLTB pool from Experiment 19 to tube E;

(b) add 700 ml PBS/1% BSA to tube F;

(c) add 700 ml PBS/1% BSA to tube G;

(d) with a fresh pipet tip, remove 700 μl of rLTB from tube E to tube F;

(e) cap the tube. Mix by vortexing;

(f) with a fresh pipet tip, remove 700 μl of rLTB from tube F to tube G;

(g) cap the tube. Mix by vortexing.

(4) Remove PBS/1% BSA blocking buffer from all plate wells by aspiration, or by flicking the solution out of the plate over the sink. Let the plate dry for 2–5 min.

(5) For the 'standard' section of the plate:

(a) add 50 μl LT CT or CTB (200 ng/ml) to each well B (1–12) to obtain 10 ng/well in row B;

(b) add 50 μl LT CT or CTB (100 ng/ml) to each well C (1–12) to obtain 5 ng/well in row C;

(c) add 50 μl LT CT or CTB (50 ng/ml) to each well D (1–12) to obtain 2.5 ng/well in row D.

	GM1			GD1a			GM2			asialo-GM1		
	1	2	3	4	5	6	7	8	9	10	11	12
A	blanks			1% BSA/PBS								→
B	←			10 ng LT/well								→
C	←			5 ng LT/well								→
D	←			2.5 ng LT/well								→
E	←			1X rLTB/well								→
F	←			0.5X rLTB/well								→
G	←			0.25X rLTB/well								→
H	←			10 ng CT/well								→

Figure 23.6 Plate plan

Standards for direct concentration determinations are best prepared by direct dilution so if one is wrong the curve is not compromised. More than three points are generally necessary. This format should afford a general estimate of concentration.

(6) For the 'sample section':

(a) add 50 µl LTB from tube 'E' into each well E (1–12);

(b) add 50 µl LTB from tube 'F' into each well F (1–12);

(c) add 50 µl LTB from tube 'G' into each well G (1–12).

(7) Control rows: A(1–12) – add 50 µl PBS/1% BSA/well = blanks; H(1–12) – add 50 µl CT/well (10 ng/well) = comparison.

For the sake of simplicity only one primary antibody (anti-CTB) will be used to probe for both LTB and CTB. The homologs bind only slightly less well to their non-cognate antibodies.

(8) Allow samples to incubate for 30 min. Previous experience [Figure 23.5(a)] suggests that binding to ganglioside ligands is very rapid, so the incubation time is short for this step.

(9) Remove samples by aspirating with a Pasteur pipette. Expel waste into a beaker containing 95% ethanol (to denature the toxic proteins).

(10) Wash three times with wash buffer; fill wells nearly full with wash buffer, then flick plate empty into the sink; repeat two more times.

(11) Add 50 µl PBS/1% BSA/well to wells A(1–12) blanks.

(12) Add 50 µl 1:5000 anti-LT antibody (primary antibody) to remaining wells (B–H, 1–12).

(13) Incubate for 45 min to allow time for antibody to bind to all available antigen, see Figure 23.5(b).

(14) Flick out antibody and PBS/BSA into the sink.

(15) Wash three times with wash buffer as described in step (10).

(16) Add PBS/1% BSA to wells A(1–12) blanks.

(17) Add 50 µl 1:5000 anti-goat IgG-HRP (secondary antibody) to remaining wells (B–H, 1–12).

(18) Incubate for 30 min.

(19) Wash three or four times to be certain no unbound or nonspecifically adsorbed peroxidase remains in the wells.

(20) Add 50 µl one-step TMB-ELISA or other soluble chromophore.

(21) Incubate for 10–20 min.

(22) Read visually or use a microplate reader (spectrophotometer). The blue color is read at 650 nm. The reaction can be stopped using 10 µl of sulfuric acid. It turns yellow and is read at 450 nm.

Preparation for experiment

(1) PBS; see Experiment 17. Any standard phosphate-buffered saline (PBS) can be used, such as 10 mM sodium phosphate/500 mM NaCl, pH 7.4, or Dulbecco's PBS (10 mM sodium–potassium phosphate/500 mM NaCl, pH 7.4; Pierce Chemical, catalog no. 28374) containing 10 ml 10% BSA/100 ml makes a very consistent formulation. Pre-mixed packets and tablets are commercially available. TBS can be substituted for PBS.

(2) PBS/1% BSA: 10% stock (5 g BSA+45 ml water), 0.2 µm filtered. Store at 4°C to reduce bacterial growth and dilute to 1% for use. Add 5 ml of 0.2 µm filtered 10% BSA to a 50 ml sterile centrifuge tube and make up to 50 ml with 0.2 µm filtered or

autoclaved distilled water. Any remaining 1% BSA should be discarded after use. It becomes contaminated very easily.

(3) Wash buffer PBS/0.1% BSA, pH 7.2 containing 0.05% Tween-20 (see Experiment 20).

(4) LT or CTB diluted to: 200 ng LT/ml PBS/1% BSA = 10 ng LT in 50 μl; 100 ng/ml PBS/1% BSA, 5 ng LT/50 μl; 50 ng/ml PBS/1% BSA, 2.5 ng LT/50 μl.

(5) 1:50 stock anti-CTB IgG diluted to 1:5000 in PBS/1% BSA (see Experiment 20).

(6) 1:50 stock anti-goat IgG-HRP conjugate diluted to 1:5000 in PBS/1% BSA immediately prior to use (see Experiment 20).

Note: 5 ml is required to fill a 96-well u bottom plate with 50 μl/well.

References

Spangler, B. D., Wilkinson, E. A., Murphy, J. T., Tyler, B. J. (2001). Comparison of the Sprecta surface plasmon resonance sensor and a quartz crystal microbalance for detection of *Escherichia coli* heat-labile enterotoxin. *Anal. Chimica Acta* **444**, 149–161.

EXPERIMENT 24

Protein Structure Determination: X-ray Diffraction Techniques

Brief introduction to X-ray crystallography

The tertiary structure of a protein is described by its folding patterns, with each atom position precisely defined in three-dimensional space. Knowledge of the structure allows an observer to literally see how each atom in a protein structure relates to each other atom, how they interact with one another and with the environment and, with that knowledge in hand, to surmise the ways in which those structural interactions contribute to the functions of the protein. Further consideration leads to speculation about how alterations in the amino acid sequence could lead to altered protein function or to the acquisition of new functions by known proteins.

Determination of the three-dimensional structure of proteins grew from small molecule single-crystal X-ray crystallography and blossomed with the advent of computer programs that handle immense amounts of data very quickly. In that same period, nuclear magnetic resonance (NMR), also the beneficiary of greatly enhanced instrumentation, contributed many dynamic structures of proteins as they occur in solution. A small discussion and figure showing a structure can be found in Voet, Voet, and Pratt (2002; on p. 142). Protein structure determination begins with a pure protein. Purified native proteins are good but often found in very small amounts. Modern molecular biology techniques allowing over-expression and cloning of homologous proteins make sizable quantities of purified recombinant protein material available, sufficient to facilitate structural studies. While small proteins can be analyzed by NMR in solution, structure determination by NMR is much more difficult for very large macromolecules. On the other hand, a good protein crystal contains many molecules of a single protein arranged in an ordered array, a repeating pattern, so that information from atoms in each molecule in the crystal reinforces the data accumulated. The structure in a protein crystal, however, is a time- and population-averaged one in which conformation may be influenced by the crystallization conditions so the structure should be viewed more as a guide to active function. In the case of proteins that retain their enzymatic activity in the crystalline form, and most do, the structure will be a good representation of the active conformation.

The goal of X-ray diffraction analysis is to obtain a detailed picture of the contents of the crystal at the atomic level, as if one had viewed it through an extremely powerful microscope. The analogy is shown in Figure 24.1(a) and (b) (Glusker and Trueblood, 1985). Both methods use scattered radiation and in both cases the object scatters some of the incident radiation into a diffraction pattern. In a light microscope the objective lens focuses the scattered light to give a magnified image of the sample. The closer this lens is to the object, the wider the angle through which scattered radiation is caught by the lens and most of the diffracted light will contribute to forming the image. With X-rays, the diffraction pattern has to be captured and recorded electronically or photographically because X rays cannot be focused by any known lens. Therefore the recombination of the diffracted radiation done by a lens in the light microscope must be done mathematically instead, by a crystallographer, with the aid of complex, efficient computer programs, although simpler small molecule structures were solved using mechanical calculators as recently as the 1970s. The recombination of diffracted radiation cannot be done directly because the phase relations among the different diffracted beams cannot usually be measured directly. The information is lost when the diffraction pattern is recorded. Once the phases have been mathematically deduced, an image of the scattering matter can be formed and the X-ray structure determined.

Why X-rays?

In order for an object to diffract light, the wavelength of the light must be approximately the same size as the object. That is why visible light cannot produce images of individual atoms: the visible electromagnetic radiation spectrum is 400–700 nm, but bonded atoms in a protein molecule are only 0.15 nm (1.5 Å) apart. In order to visualize atoms this close together, electromagnetic radiation in the X-ray range is required. Nevertheless, hydrogen atoms are too small to be visualized (although water molecules can be), so the

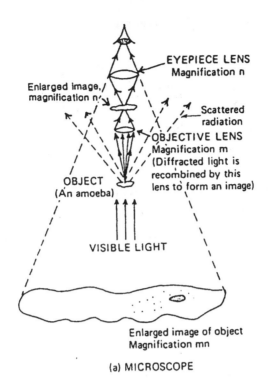

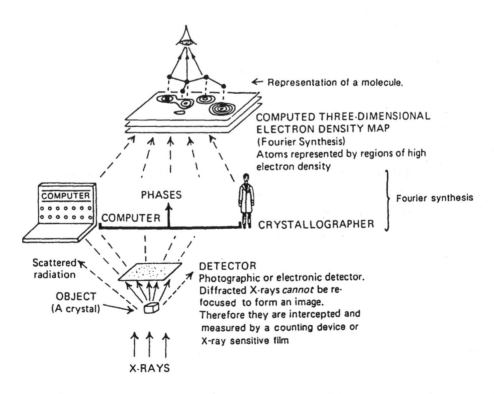

Figure 24.1 (a) Light microscope; (b) X-ray diffraction (Glusker and Trueblood, 1985). Reproduced by permission of Oxford University Press

H atoms in a protein structure are added, assuming that bond angles are the same as in small molecules.

A single molecule is a weak scatterer of X-rays. Most X-rays will pass through a single molecule without being diffracted, leaving the diffracted beam too weak to be detected. As previously noted, crystals are ordered arrays of molecules so each molecule diffracts identically and the diffracted beams therefore augment each other to produce strong detectable patterns on an X-ray film, or a diode array detector (Rhodes, 2000). The larger the crystal, then, the more clearly the paterns can be detected. Crystals of 0.5 mm are a very good size, although smaller ones can normally be analyzed without trouble using current instruments. The degree of order within the crystal is another consideration, since disordered arrays result in destructive interference and therefore reduced diffraction.

Crystals

The internal periodicity of crystals was hypothesized based on the regularity of crystal shapes and was demonstrated in 1912 when it was shown that a small molecule crystal could act as a three-dimensional diffraction grating for X-rays. The basic building block for a crystal is the unit cell. The crystal lattice is a three-dimensional array of points upon which the contents of the unit cell are arranged in infinite repetition to build the structure of a crystal. There are only 14 distinct kinds of lattice, called the Bravais lattices, that may be constructed such that the view from each lattice point is the same in a given direction as the view in that direction from any other lattice point. These lattices have seven different unit cell shapes with different symmetry properties, which correspond to seven crystal systems. The arrangement of structures, molecules, on these lattices must be consistent with one of the 230 different combinations of symmetry elements, known as space groups, that are possible for arranging objects in a regularly repeating manner in three dimensions (Glusker and Trueblood, 1985). This topic is well-described in several books, most readably by Rhodes (1985), and by Glusker and Trueblood (2000). Protein crystals have a much more limited subset of possible arrangements (a total of 64).

Diffraction

In crystals, the electrons in the atoms act as scatterers for X-rays. The diffraction pattern of a crystal, displayed as spots on an X-ray film, are arranged on a lattice that is reciprocal to the lattice of the crystal. In practice, the larger the unit cell the closer the

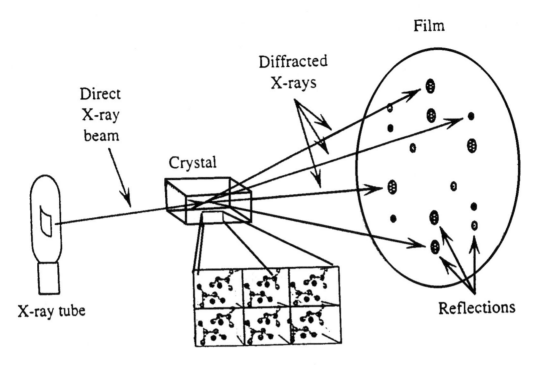

Figure 24.2   Crystallographic data collection. Reproduced by permission of Academic Press

reflections will be to one another. Figure 24.2. indicates the reciprocal nature of the process.

The detected spots are called reflections, as they represent the positions of the diffracted beams, reflection from planes through points in the crystal lattice. Structure analysis by diffraction methods can be understood as analogous to diffraction by gratings, extended to three dimensions. The positions and intensities of the reflections contain the information needed to determine molecular structures.

Since X-rays interact with electrons in matter, rather than the atomic nuclei, the X-ray structure, when resolved, is an image of the electron density of the sample. The position of the reflections depends on dimensions of the crystal lattice, while the intensities of the different diffracted beams depend on the nature and arrangement of the atoms within each unit cell. This information can be shown as a three-dimensional contour map [see Figures 6-22 and 6-23 in Voet, Voet and Pratt (2002, pp. 140–142)]. When these factors are determined, the relationship of atoms to one another, the structure of the sample, can be deduced. The determination depends on an analysis of X-ray waves scattered from different points in a given direction. The intensity scattered at any angle can be calculated from the sum of the waves scattered from all the electrons within the unit cell. The structure determination requires the observed intensity pattern to be matched to a postulated model. The model represents the approximate positions of the atoms in the sample, usually based on the primary structure of the protein, and is required in order to obtain the phase of the scattered waves, a crucial variable in the equations that describe shapes in three dimensions (the Fourier transform). Phase information is lost when the X-rays impinge on the detector film or diode array. The phase of the X-ray wave must be mathematically deduced from information in a model, then used as a variable in the Fourier transform that will describe the sample. In effect, the Fourier transform equations become the lens that allows computation of electron density within the unit cell.

Data collection

The goal of data collection is to determine the positions and intensities of as many reflections as possible in as short a time as possible. Timing is important because bombarding protein crystals with X-rays causes deterioration of the protein due to heat generation and the production of reactive free radicals by the ionizing radiation. In a matter of hours a protein can turn into 'rubber', losing its conformation and its ability to diffract the X-rays. Freezing the crystal is one method for reducing these hazards.

To collect data, a protein crystal is sealed into a very fine capillary tube, usually in the presence of a bit of the crystallizing liquor from the reservoir. This process keeps the crystal in a moist atmosphere and 'preserves' it. Alternatively, the crystal is flash-frozen. The crystal is then attached to a complex rotational device called a goniometer head which can rotate and translate the crystal in the beam in very precise directions. The rotation is accomplished by computer control. Frozen crystals are maintained in a stream of liquid nitrogen. The goniometer is mounted onto the diffractometer, which directs single reflections to a scintillation counter or an area detector, or several cameras capable of directing a large number of reflections to film or area detectors. In addition, the diffractometer has instrumentation capable of moving the goniometer head, so that the crystal can be irradiated in almost any direction. The diffractometer also includes a fixed X-ray source. X-rays are electromagnetic radiation in the range 0.1–100 Å. X-rays at wavelengths useful for protein crystallographic applications are produced by bombarding a metal target, usually copper with electrons produced by a heated filament and then accelerated by an electric field. The process is accomplished when an accelerated, high-energy electron collides with and displaces an electron from a lower orbital in an atom of the target metal. Then an electron from a higher orbital drops in the resulting vacancy, and in so doing, emits its excess energy as an X-ray photon.

There are three common X-ray sources: cathode ray tubes, such as those found in a television; rotating anodes; and particle storage rings. The particle storage rings produce synchrotron radiation when electrons are accelerated around a ring and loose energy in

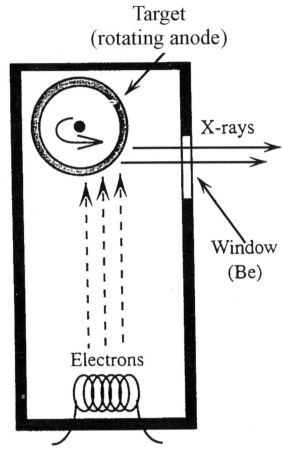

Figure 24.3   Data collection (Rhodes, 2000). Reproduced by permission of Academic Press

the X-ray region as they turn corners. Since the tubes generate heat and must be cooled by water, their output is limited. Diffractometers in general use are usually powered by a rotating anode. These are the work-horse instruments found in macromolecular X-ray crystallography laboratories. Synchrotron radiation is much more intense, so shorter exposure times are possible, seconds rather than hours, but the rings are found in only a few sites, so travel to a synchrotron radiation source is necessary. The X-ray beam is directed through a narrow metal tube called a collimator, that selects and reflects the X-rays into a narrow parallel beam. Metal plates are used to further narrow the beam so that it will penetrate the crystal in a very small area.

The resulting reflections can be measured in a variety of ways. Initially, X-rays were measured on sensitive film. The intensity was determined by a densitometer. A scintillation counter counts X-ray photons and provides accurate intensities. The counter contains a material that produces a flash of light, hence scintillation, when it absorbs an X-ray photon. A photocell counts the flashes and the number is recorded. The film technique has been replaced with image detectors and charge-coupled devices, CCD cameras that direct reflections to detectors for data collection.

Data analysis and structure determination

Data refinement, manipulation and the transformation of the data collected into an electron density map require knowledge of the mathematical concepts required to describe a unit cell and symmetry consideration, which go into the preparation of a Fourier transform. Many of these manipulations can be done using a computer program, but a good crystallographer must always understand the nature and composition of the variables that are manipulated by the computer program. It is beyond the scope of this exercise to detail them, but a number of references in addition to the cited works will prove informative for those students interested in three-dimensional structure determination.

References

Glusker, J. P. and Trueblood, K. N. (1985). *Crystal Structure Analysis: a Primer*, p. 262. New York: Oxford University Press.
Rhodes, G. (2000). *Crystallography Made Crystal Clear*, p. 267. San Diego, CA: Academic Press.
Voet, D., Voet, J. G. and Pratt, C.W. (2002). *Fundamentals of Biochemistry*, p. 931. New York: John Wiley.

EXPERIMENT 25

Characterization of a Protein Crystal using X-ray Diffractometry (optional)

This exercise is dependent on the availability of a Protein Crystallography Laboratory. It is left as an optional exercise, depending on users' circumstances. The crystals of lysozyme produced in Experiment 21 are generally of excellent quality for the purpose. Crystals of LTB may be more problematic but, if available, are most interesting.

Alternatively, if a small molecule crystallographer is available, that person would be an excellent resource for describing the principles of crystallography and structure determination.

EXPERIMENT 26

Primer Design

Introduction

The goal of this entire experimental project is to obtain protein for both structural and functional studies, and to make use of these studies to understand the mechanisms for molecular recognition. Elucidating these mechanisms should lead to design of improved affinity and broadened specificities for the LTB pentamer. X-ray studies have demonstrated that the pentameric structure is required for CTB (and LTB) to bind its receptor, ganglioside GM_1 (Merritt *et al.*, 1994). Figure 24.1(a) and (b) makes it clear that the binding pocket is the interface between monomers, and binding is mediated by Gly33 from one monomer and Trp88 on the adjacent monomer. This information was deduced by site-specific mutation (Jobling and Holmes, 1991) and a collection of earlier studies reviewed (Spangler, 1992) but the X-ray data make it incontrovertible. Later studies have expanded on this theme (Merritt *et al.*, 1997).

The aim of this laboratory exercise is to use the primer design software to interactively design a set of primers that will be used to introduce a single mutation into the binding site of LTB. The amino acid to be modified depends on the proposed application for the mutant protein. There have been numerous studies leading to vaccine development based on the ability of LTB to elicit mucosal immunity, coupled with its ability to bind mucosal cells. Expansion of those characteristics, based on targeted mutations that enhance adjuvant activity and also make use of the pentamer as a delivery vehicle for a variety of immunogenic molecules, makes genetic engineering of the pentamer an attractive goal. The research articles can be found on MedLine (www.ncbi.nlm.nih.gov/entrez/query)

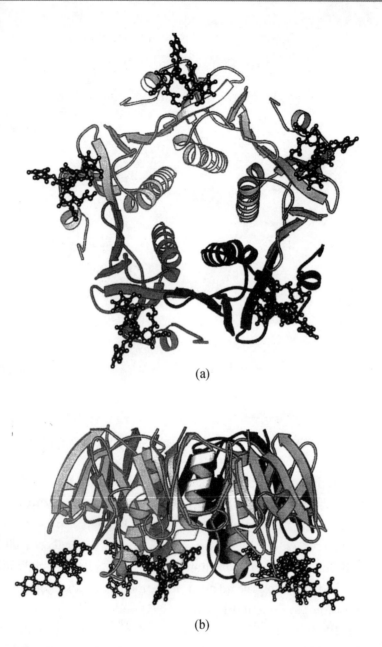

(a)

(b)

Figure 26.1   Ribbon diagram of B subunit from Merritt *et al.* (1994, access no. 604; *Protein Sci.* **3**, 166–175, Figure 3). Reproduced by permission of The Protein Society

by entering '*Escherichia coli* heat-labile enterotoxin' in the subject line. The titles that are displayed can be examined. Choose a title that describes an interesting application. Click on 'Related Articles' and choose several recent articles to read and consider. From that information, a novel application can be devised, based on mutations introduced into LTB. The result should be a continuing research project using the baseline data and protocols in the previous 25 experiments.

Outline for disussion

Considerations for effective primer design

(1) General considerations: what do you need it for?

◆ PCR (a pair of primers);

◆ mutagenesis (can be either one or two);

◆ sequencing (one).

(2) Length of primer:

◆ 18–30-mer for PCR;

◆ 20–25-mer for sequencing;

◆ 40–50-mer for mutagenesis.

(3) Target melting temperature:

◆ methods of estimating melting temperature;

◆ what it should be – 55–75°C for PCR or 55–65°C for sequencing?

(4) What will affect the real melting temperature?

◆ Mg^{2+};

◆ DMSO;

◆ formamide;

◆ glycerol.

(5) Structural characteristics of a primer:

◆ GC content;

◆ formation of hairpins;

◆ runs;

◆ 5′–3′ stability;

◆ self–self dimer formation.

(6) When considering a pair of PCR primers:

◆ close melting temperatures;

◆ formation of heteroduplexes;

◆ size of the expected product.

(7)   Are your primers 'facing each other'?

(8)   Purification of primers:

♦   standard (desalting) – for regular PCR primers;

♦   reverse-phase HPLC – optional for sequencing primers;

♦   PAGE purification – a must for mutagenic and labeled primers (generally recommended for primers longer than 30 bases).

(9)   Special cases:

♦ $5'$-extentions;

♦   leaving a spacer at the $5'$ terminus;

♦   first round and second round melting temperatures;

♦   derivatization of primers: – $5'$-dephosphorylation, biotinylation, fluorescent labeling, digoxygenin labeling, attach almost anything.

(10) Final note: cloning in frame.

Materials

♦   On the Web: search 'DNA primer design' to find numerous entries, some free

♦   See Web Primer (Stanford): http://genome-wwwz.stanford.edu/cgi-bin/SGD/web-primer and www.alkami.com/primers/refdsgn.htm for additional information

References

Jobling, M. G. and Holmes, R. K. (1991). Analysis of structure and function of the B subunit of cholera toxin by the use of site-directed mutagenesis. *Mol. Microbiol.* **5**, 1775–1767.

Merritt, E. A., Sarfaty, S., Akker, F. v. d., L'Hoir, C., Martial, J. A. and Hol, W. G. J. (1994). Crystal structure of cholera toxin B-pentamer bound to receptor GM_1 pentasaccharide. *Protein Sci.* **3**, 166–175.

Merritt, E. A., Sarfaty, S., Jobling, M. G., Chang, T., Holmes, R. K., Hirst, T. R. and Hol, W. G. (1997). Structural studies of receptor binding by cholera toxin mutants. *Protein Sci.* **6**, 1516–1528.

Rhodes, G. (2000). *Crystallography Made Crystal Clear*, p. 267. San Diego, CA: Academic Press.

Spangler, B. D. (1992). Structure and function of cholera toxin and the related *Escherichia coli* heat-labile enterotoxin. *Microbiol. Rev.* **56**, 6.

Index